T0324700

Springer Undergraduate Texts in Philosophy

[illegible] textbooks aimed towards the undergraduate level that covers all areas of philosophy ranging from classical philosophy to contemporary topics in the field. The texts will include teaching aids (such as exercises and summaries) and will be aimed mainly towards more advanced undergraduate students of philosophy.

The series publishes:

- All of the philosophical traditions
- Introduction books with a focus on including introduction books for specific topics such as logic, epistemology, German philosophy etc.
- Interdisciplinary introductions – where philosophy overlaps with other scientific or practical areas

This series covers textbooks for all undergraduate levels in philosophy particularly those interested in introductions to specific philosophy topics.

We aim to make a first decision within 1 month of submission. In case of a positive first decision the work will be provisionally contracted: the final decision about publication will depend upon the result of the anonymous peer review of the complete manuscript. We aim to have the complete work peer-reviewed within 3 months of submission.

Proposals should include:

- A short synopsis of the work or the introduction chapter
- The proposed Table of Contents
- CV of the lead author(s)
- List of courses for possible course adoption

The series discourages the submission of manuscripts that are below 65,000 words in length.

For inquiries and submissions of proposals, authors can contact Christi.Lue@springer.com

More information about this series at http://www.springer.com/series/13798

The Springer Undergraduate Texts in Philosophy offers a series of self-contained textbooks aimed towards the undergraduate level that covers all areas of philosophy ranging from classical philosophy to contemporary topics in the field. The texts will include teaching aids (such as exercises and summaries) and will be aimed mainly towards more advanced undergraduate students of philosophy.

The series publishes:

- All of the philosophical traditions
- Introduction books with a focus on including introduction books for specific topics such as logic, epistemology, German philosophy etc.
- Interdisciplinary introductions – where philosophy overlaps with other scientific or practical areas

This series covers textbooks for all undergraduate levels in philosophy particularly those interested in introductions to specific philosophy topics.

We aim to make a first decision within 1 month of submission. In case of a positive first decision the work will be provisionally contracted: the final decision about publication will depend upon the result of the anonymous peer review of the complete manuscript. We aim to have the complete work peer-reviewed within 3 months of submission.

Proposals should include:

- A short synopsis of the work or the introduction chapter
- The proposed Table of Contents
- CV of the lead author(s)
- List of courses for possible course adoption

The series discourages the submission of manuscripts that are below 65,000 words in length.

For inquiries and submissions of proposals, authors can contact Christi.Lue @ springer.com

More information about this series at http://www.springer.com/series/13798

Lars-Göran Johansson

Philosophy of Science for Scientists

Lars-Göran Johansson
Filosofiska Institutionen
Uppsala Universitet
Uppsala, Sweden

Figure 3.1 reproduced with kind permission of Tunc Tercel.
Figure 7.2 reproduced with kind permission of Lou-Lou Pettersson.

Springer Undergraduate Texts in Philosophy
ISBN 978-3-319-26549-0 ISBN 978-3-319-26551-3 (eBook)
DOI 10.1007/978-3-319-26551-3

Library of Congress Control Number: 2015958786

Springer Cham Heidelberg New York Dordrecht London

Printed on acid-free paper

Springer International Publishing AG Switzerland is part of Springer Science+Business Media (www.springer.com)

Preface and Overview of the Book

The Swedish predecessor of this book, *Introduktion till Vetenskapsteorin*, grew out of an urgently felt need when I was teaching philosophy of science for students of engineering, physics, biology, social science, medicine and nursing. These students have normally no philosophical background and quite often little knowledge of history of science. This book has now been in print for 15 years, and three editions and its relative success in Sweden have encouraged me to make a translation to English in the hope that a wider audience also will find it useful.

This book is not *merely* a translation of the Swedish book; I have also made some changes. First, Ties Niessen suggested a slight reshuffle of the chapters and an addition of a short Chap. 14, with some actual and forward-looking reflections, which I have done. Second, I have rewritten Sect. 10.7, since I have come to understand laws better. Third, I have made a great number of minor changes as a result of comments and suggestions from two anonymous referees. Their advice was very helpful.

The prime goal for a first course in philosophy of science should be, I believe, to convey an understanding of what science is: how it has developed, what its core traits are, how to distinguish between science and pseudoscience and to know what a scientific attitude is.

In such an endeavour it is common and natural to concentrate on the development and core traits of natural science. However, students and scholars within the social sciences and humanities often think that these sciences differ profoundly from natural science and that the lessons from Galilei, Newton and other natural scientists are not relevant for them.

Here a remark about the word 'humanities' is in place. Hume and other eighteenth-century British philosophers used the word 'moral sciences' as a label for studies we now would call 'humanities'. The effect is that the word 'science' without modifier now means natural science only. This is not so in German, Swedish and other Germanistic languages, where the corresponding words ('Wissenshaft' 'vetenskap') are used for all systematic studies at universities. It seems to me that using the word 'humanities' encourages people to see the differences rather than the similarities among different disciplines, and since I want to stress commonalities among the sciences, I suggest using the expression 'human sciences' as replacing 'humanities'.

It is commonly assumed that natural science is concerned with testing hypotheses and discovering natural laws, whereas the aim of human and some social sciences typically is to achieve understanding, i.e. understanding *the meanings* of individual's and social group's actions. Such understanding may be achieved by some interpretative method, which is seen as profoundly different from the method of testing hypotheses.

I have no objections against these two broad characterisations of respectively natural science and human and social sciences, but I disagree about the tacit assumption that testing of hypotheses and making interpretations – doing hermeneutics – are radically different activities. In fact, I think a good case can be made for the view that interpretation of texts, utterances, behaviour, cultural phenomena, etc., are species of hypothesis testing, not, of course, hypotheses about regularities, as in natural and some social sciences, but about *meanings*. The structural similarities between the hypothetico-deductive method, the hermeneutic circle and Davidson's rules for interpretation are not difficult to recognize, once one has freed oneself from the idea that hypotheses by necessity are about regularities in the world. Dagfinn Føllesdal was, as far as I know, the first to point out these similarities. However, it is still a controversial view and one aim of the book is to give some arguments for it.

But why stressing similarities between the sciences? The main reason is that we need to say something general about all sciences in order to effectively demarcate between science and pseudoscience, which in my view is a prime duty when teaching elementary philosophy of science. Pseudoscience is quite popular and many people are astonishingly credulous and/or prey of wishful thinking. And some people just dress up their activities, whatever they are, by calling them 'science' just because it enhances the prestige of what they are doing, or so they think.

So how to demarcate? It won't do to say that each particular science has its own rules of inquiry, its own criteria for being scientific, because then proponents of, for example, homeopathy could say: 'Yes, we agree that every discipline has its own criteria and that applies as well to homeopathy: our criteria differ somewhat from school medicine (an expression often used by homeopaths when they talk about medical science taught at universities) but our criteria are just as scientific as those of school medicine and we are just as scientific as they are. Proponents from school medicine act as imperialists on the market for theories about treatment of diseases when they denounce us'.

This argument we need to rebut, and the way to do that is to argue that scientific thinking, independent of domain of inquiry, ought to satisfy some fundamental and general epistemic demands. Hence, we need common criteria for any activity properly being called scientific. I believe that the hypothetico-deductive method and strictures on valid observation reports are the main components in such a list of criteria.

Some people are sceptical about the possibility of finding general criteria for science, although they see the need. In particular, it has been argued that the hypothetico-deductive method is too strong a criterion for scientific work, since there are some activities that best are described as 'data mining' or 'data collection' in some scientific disciplines, activities that are not driven by any explicit hypothesis, and we do not want to dismiss such activities as unscientific. I agree that we do not want to do that. However, hypothetico-deductive method is not a criterion for *every* activity called 'research' in a discipline; it is better viewed as a criterion on the discipline as a whole. The fact that some researchers in some disciplines sometimes engage is 'data mining' or the like doesn't entail that the discipline fails the general criterion for being a science.

The need for general criteria for scientific thinking is no more than an instance of the epistemological demand to produce reasons, acceptable to others, for your claims to know. Rational scientific discussions about methods, measurements, inferences and conclusions presuppose that it is possible to discuss and agree on epistemological and scientific norms independently of whether one accepts the conclusions of a particular theory or not. It won't do to have acceptance of the method used from only those who already believe the theory and its results. (There is indeed a problem here; some areas of research such as advanced mathematics or modern theoretical physics are understood by a very limited number of researchers, but I leave that aside for the moment.)

Hence, I believe it is very important to have some sort of general conception of all sciences when discussing the demarcation between science and pseudoscience.

A related topic is the theory-relatedness of observations; some have claimed that there are no such things as fully theory-independent observations. If true, it would undermine the possibility of objectivity of science and force us to accept strong relativism. I believe that this disastrous consequence can be avoided and that there really is a basis of theory-neutral data, also in the humanities. This is the topic of Chaps. 4 and 5.

These considerations have guided the structure of the first part of the book, consisting of Chaps. 1, 2, 3, 4, 5 and 6.

The second part consists of Chaps. 7, 8, 9, 10 and 11. In these chapters I discuss topics I have found relevant and useful to talk about even at an introductory philosophy of science course, viz. *causes, explanations, laws and models*. Causation is arguably the most important of these topics since almost all empirical disciplines contain causal idiom to some extent and the search for causes is in many disciplines a prime goal.

The notion of explanation is often connected to causation, but the use of the word 'explanation' differs profoundly from context to context and one may wonder if there really is anything in common to everything we call an 'explanation'. This is the topic of Chaps. 8 and 9.

Laws and models are core concepts in natural sciences but less so in social science and perhaps not at all in human science. The discussion about the concept of natural law is intense among philosophers of science and a lot of views have been propounded. In Chap. 10 I discuss some of them and indicate my own empiricist position.

By contrast, models are not much discussed among philosophers of science. This is a bit astonishing, since scientists very often talk about models when discussing the 'fit' between theory and reality. One is immediately prone to ask what kind of epistemological and ontological status models have. In Chap. 11 I discuss this and what scientists might mean with their talk about models.

The final part consists of some additional material that is naturally brought up in a philosophy of science course, although it does not belong to philosophy of science proper.

Chapter 12 is about some issues in philosophy of mind, a topic usually not covered in a philosophy of science course, or book. The reason I nevertheless have included a brief discussion about mind states is that in particular students in psychology and medicine are naturally confronted with difficult questions about the relations between mind and body. My experience is that these students somewhat unreflectively adopt a vocabulary reflecting substance dualism, for example, the distinction between biological and psychological causes of mental diseases and aberrant behaviour. (And neurophysiology and psychiatry are considered as two different medical subdisciplines, a distinction suggesting a traditional mind-body dualism.) However, when asked about what they think about the matter, most are prepared to say that the mind and the brain somehow are identical or two sides of the same coin. In short, their position is unstable and needs to be discussed.

Chapter 13 contains a discussion of some aspects of values in science, the most important being the discussion about the concepts *value-free* and *value-laden*, once introduced by Weber. The important point is that science is driven by values, it is value-laden, but its results can, and should, be value-free.

Chapter 14 contains some reflections on actual trends in science. It is forward-looking and much more tentative than the rest of the book.

Finally, there is a short appendix about logical form, which hopefully can be useful in discussions of hypotheses testing, and in some accounts of explanation. I have in particular experienced student's difficulties in understanding the truth conditions of the material conditional, a topic which hardly can be avoided when analysing hypothesis testing. It seems to me that students without logical training often interpret conditional statements in their context as either causal or logical statements.

Thus, the reader I have had in mind is first and foremost a student taking a course in philosophy of science without having studied philosophy earlier.

The book is also useful as textbook for an introductory course at undergraduate level for students majoring in philosophy. A number of colleagues, and myself, have used the Swedish predecessor of this book in such courses and our experience is that it is well suited also for that purpose. In such courses we normally omit Chap. 2 (knowledge) and Chap. 12 (philosophy of mind), since these topics are covered in other philosophy courses.

Two of my former students, David McVicker and George Masterton, both of which are native English speakers, have helped me with the English translation of my Swedish textbook. David did the first draft, which then was checked by George.

The result is much better than what I could have done myself. My gratitude is hereby acknowledged.

Two anonymous referees for this English version have given many valuable comments, which hereby is gratefully acknowledged. Finally, I thank Ties Nijssen for much help and encouragement in the final editing of the book.

Uppsala, Sweden Lars-Göran Johansson
summer 2015

Contents

Part I
What Is Science?

Chapter 1
The Evolution of Science

And which of the gods was it that set them on to quarrel? It was the son of Jove and Leto; for he was angry with the king and sent a pestilence upon the host to plague the people, because the son of Atreus had dishonoured Chryses, his priest. Now Chryses had come to the ships of the Achaeans to free his daughter, and had brought with him a great ransom: moreover he bore in his hand the sceptre of Apollo wreathed with a suppliant's wreath and he besought the Achaeans, but most of all the two sons of Atreus, who were their chiefs.
Homer: Iliad

1.1 Greece: The Dawn of Science

Only a short moment's contemplation shows us that science plays an enormous role in modern western societies. Scientific achievements have improved quality of life for billions of people in many ways. On the other hand, in certain respects, scientific advancements have worsened our quality of life. In addition, scientific advancement has often come at the expense of the rest of nature. In fact, some would even argue that science causes more harm than good, an opinion which I deem absurd. In any case, everyone can agree that science has greatly influenced the course of life all over the world.

From a global perspective, the evolution of science, and the consequent transformation of society, is a unique process. Throughout the ages, many societies have attained high levels of cultural development and organization (streets and roads, irrigation, educational institutions, organized postal service, formal justice systems, etc.). Yet, for the most part, modern western societies were the first to develop systems that saved large portions of its population from starvation and lengthened the average lifespan. Natural questions one might wish to ask in the light of these phenomena are how this societal change began, why it began in Western Europe, and why it began when it did?

Most would agree that these questions regard a profound change–a true revolution–that is considered the most significant historical development since the invention of agriculture. One can distinguish two somewhat distinct processes in this revolution: (i) the scientific revolution, which is usually placed in the period 1550–1650 A.D. and (ii) the industrial revolution, which began in England in the

L.-G. Johansson, *Philosophy of Science for Scientists*, Springer Undergraduate Texts in Philosophy, DOI 10.1007/978-3-319-26551-3_1

1700s. The connection between these two revolutions is a controversial issue: some think that the industrial revolution was, for the most part, independent of the science of the day, while others think that a pertinent precondition for this process is the evolution of science and the so-called *scientific perspective*.

That the industrial transformation of western societies in the last hundred years is largely a result of scientific developments is undisputed. However, there is a question as to the role played by scientific insights in the beginning of the industrial transformation. In other words, we may ask, when did science begin to gain industrial and economic importance? Without taking a definite stance to this question, one can point at Sadi Carnot's (1796–1832) studies concerning the effectiveness of steam engines as a clear example of a scientific study, which was not founded on pragmatic demands, and yet had direct technical relevance. In his famous book 'Reflections on the Motive Power of Fire' (1824) Carnot discusses the effectiveness of the steam engine, starting from purely theoretical principles. Guided by his insights, one could immediately introduce substantial improvements in the efficiency of steam engines. A crucial step was Carnot's introduction of the concept of a quantitative measure of heat, *Calorie*. The previous long-lived idea that heat was a substance was replaced by the conception of heat as a quantitative property and the successful use of Carnot's theory convinced people that this new conception was correct.

However, the scientific revolution started much earlier at around the middle of the 1500s in northern Italy, England, and Holland, the commercial centres of the time. Why then and in those areas? What are the causes of the scientific revolution? These questions are not only of historical interest! Many developing countries are now quickly attempting to embrace scientific and technological achievements with the kind assistance of humanitarian an U.N. organizations. However, the results of these endeavours are varied and sometimes discouraging. Thus there is anecdotal evidence that we lack sufficient understanding of the essential aspects of this process of transition.

In all likelihood, a wide range of circumstances is required for scientific thinking to emerge in a society and be applied in economic and industrial contexts. Among these circumstances one may distinguish between *external* and *internal* factors. External factors include, but are not limited to, religious, economical and political factors, and they act in fairly obvious ways as external constraints on our society's intellectual development. Internal factors include theories of knowledge, world-views, and norms of argumentation, which are, by and large, inherited. The ideas and concepts inherited from ancient thinkers is a factor that I shall argue actively contributed to the scientific revolution gaining momentum during the renaissance in Western Europe. We must keep in mind that many cultures have preceded ours in history; and yet, none of them achieved a similar breakthrough in scientific thinking. Which components of this ancient heritage were significant, and what was lacking from those earlier cultures whose absence prevented them from developing an active science?

As every history must begin somewhere, and as the germ of science is clearly gleaned in ancient Greece, let us begin this brief study of the historical preconditions of science there. To fully appreciate ancient Greek efforts towards scientific thinking, one would have to consider and discuss a great many things, but I shall here limit myself to four ideas that have proven particularly significant. I do not dare argue that these four ideas are the *most* significant as regards the evolution of scientific thinking, only that they are important. The four ideas are (1) the Ionian natural philosopher's way of explaining nature, (2) the emphasis on rational argumentation, (3) Aristotle's introduction of the concept of logical validity, and (4) Euclid's axiomatic mathematics.

1. Ionian Natural Philosophy We usually consider the dawn of philosophy and science to have occurred around 600 B.C. in the Greek cities along the coast of Asia Minor, particularly Miletus. The first of the famous philosophers of this period are *Thales* (approx. 600 B.C.), *Anaximander*, Anaximander (610–546 B.C.), and *Anaximenes* (?–525 B.C.). Based on the fragments left behind, and the comments of their successors, we can conclude that these three philosophers asked the question: What is it that comprises all things? What is the underlying substrate? They thought that there must be something constant lying behind all of nature's transformations: plants and animals grow, thrive, produce offspring and die, weather varies and everything in nature changes. But changes presuppose something that does not change.

Thales thought that the underlying substrate, or principle,[1] that which is responsible for the change in all things, is water. This is perhaps not so strange an idea, since all biological processes require water, and all living things contain large amounts of water. Thales perhaps reached this conclusion by noticing that a living organism requires a large amount of water for its survival, and similarly, when it dies it releases a great deal of water.

Anaximander argued that the underlying substance was not any of the then recognized elements, earth, water, air or fire, but a more primary substrate, *apeiron* ('the indefinite'), which was thought to be boundless, eternal and unchanging. His argument against Thales is quite interesting: If water was the underlying substance, then over time it would have won the cosmic battle against the other elements, and everything would have returned to water. However, this is not the case. Thus the underlying substrate must be neutral with respect to the four elements. Obviously, the tacit assumption is that changes are to be conceived as a kind of struggle between the elements.

Lastly, Anaximenes claimed that the underlying substrate or principle was air. His idea was that all the other elements were comprised of condensed or rarefied air. This theory, although false, is interesting as it is purely physical and explains the generation of the other elements.

[1] The Greek word is 'aitia' which is translated as 'cause', 'reason', or 'responsibility'.

These primitive ideas do not strike us as particularly scientific, but we must not judge them based on what we now know. Rather, they should be judged with respect to the general presuppositions and ways of thinking that existed during that time period. In this context, it is apparent that the Ionian thinkers had taken a significant step toward developing a scientific mode of enquiry. For how did the Greeks and other peoples of the time think about natural phenomena? What did one see as the cause of nature's changes? One could summarize the typical answer in one word: myths. Natural phenomena – including personal destinies – were controlled by various gods, spirits, or demons, whose favour one was want to win. One can find examples of this thinking in Homer's *Iliad* and *Odyssey* (ca. 700 B.C.). Particularly illuminating is Homer's explanation (in the beginning of *Iliad,* see the quotation at the beginning of this chapter) of an outbreak of plague in the Greek army that was besieging Troy. The cause of this outbreak was said to be that the god Apollo was angry with Agamemnon, because he, the king and commander, had insulted Apollo's priest; and therefore Apollo sent the plague. In short, changes in nature were thought to depend upon the will of the gods. That is to say, one explained natural phenomena in the same terms as one would explain human actions. The Ionian philosopher's broke with this tradition of thought; and in so doing, took a step towards a scientific approach.

A clear example of the Ionian philosopher's influence on Greek culture is Hippocrates' (460–377 B.C.) view on the origins of disease. He writes the following regarding epilepsy, called 'the sacred disease' because it commonly was thought to have been sent by the gods:

> It is thus with regard to the disease called Sacred: it appears to be nowise more divine nor more sacred than other diseases, but has a natural cause from the originates like other affections. Men regard its nature and cause as divine from ignorance and wonder, because it is not at all like other diseases. And this notion of its divinity is kept up by their inability to comprehend it, and the simplicity of the mode by which it is cured, for men are freed from it by purifications and incantations...
>
> And the disease called the Sacred arises from causes as the others, namely, those things, which enter and quit the body, such as cold, the sun, and the winds, which are never changing and are never at rest. And theses things are divine, so that there is no necessity for making a distinction, and holding this disease to be more divine than the others, but all are divine, and all human.[2]

2. Rational Argumentation Ancient Athenians took a lively interest in argumentation. This may be connected to the fact that its democratic constitution. Athens was under the classical period (ca. 400 B.C.) a democracy, although one that excluded women and slaves. Political decisions were taken by the citizens assembled at the agora. In a democracy each person who wishes to implement his ideas must convince others by arguing for them. The natural question then is to which forms of argument are effective, and thus arose an interest in rhetoric and argumentation. Another effect of democracy is a change in attitude towards other people, since one

[2] Hippocrates: *On the Sacred Disease*, http://classics.mit.edu/Hippocrates/sacred.html

cannot usually convince people by treating them as subordinates. This realization lead free male adults in ancient Athens to treat each other as equals.

The difference between ancient Athens and other highly developed cultures of the same time period, e.g. China, is illuminating. During this period, China was split into many different states, all controlled by autocratic kings. Political influence was thus only possible for the advisors to these kings. This had the effect that more authoritarian attitudes were promoted. In order to illustrate the difference between Chinese and Greek societies during this period, we can compare two texts. The Chinese text is taken from 'Analects' by Confucius, or Kung Fu Tzu (551–479 B. C.), in which he (who made his living as an advisor to kings and rulers) is conversing with his students. The Greek text is Plato's 'Theaetetus', in which Socrates is conversing with Theaetetus, a young man who is studying mathematics:

From 'Analects':

Tzu Kung (one of Confucius' disciples) asked about the conditions for government.

The Master replied	'The requisites for the exercise of power are enough food, enough weapons, and the confidence of the people.'
Tzu Kung said	'Suppose you had to do without one of these; which would you give up first?'
The Master said	'Weapons.'
Tzu Kung said	'What if you had to give up one of the remaining two; which would it be?'
The Master said	'Food. All men must die, but a state cannot survive without the confidence of its people.[3]'

From 'Theaetetus'

Theaetetus	Well, Socrates, after such encouragement from *you*, it would hardly be decent for anyone not to try his hardest to say what he has in him. Very well then. It seems to me that a man who knows something perceives what he knows, and the way it appears at present, at any rate, is that knowledge is simply perception.
Socrates	There's a good frank answer, my son. That's the way to speak one's mind. But come now, let us look at this thing together, and see whether what we have here is really fertile or a mere wind-egg. You hold that knowledge is perception?
Theaetetus	Yes.

[3] Different translations have divided Analects differently. At http://classics.mit.edu/Confucius/analects.3.3.html the quotation is in section 3, part 12, at http://www.indiana.edu/%7Ep374/Analects_of_Confucius_(Eno-2015).pdf the passage is found in book 12.

Socrates	But look here, this is no ordinary account of knowledge you've come out with: it's what Protagoras used to maintain. He said the very same thing; only he put it rather a different way. For he says, you know, that 'Man is the measure of all things: of the things which are, that they are, and of things which are not, that they are not.' You have read this, of course?
Theaetetus	Yes, often.
Socrates	Then you know that he puts it something like this, that as each things appears to me, so it is for me, and as it appears to you, so it is for you – you and I each being a man?
Theaetetus	Yes, that is what he says.' (Trans. M. J. Levett)

The exemplified difference in attitude between teacher and pupil is pretty clear, in my view.

3. The Discovery of Logic *Aristotle* (384–322 B.C.) made many important contributions to science and philosophy, but we shall here only discuss his introduction and analysis of *logical validity*. Whether an argument is valid has to do with its *form*, and not with the *meaning* of the words used–excepting the logical ones–used in the argument. Aristotle was the first to make this distinction, and the first to study the general rules for argumentation. Here I shall briefly discuss his theory of *syllogisms*. A syllogism is an argument that can be constructed using the following four sentence types:

All A are B.
Some A are B.
Some A are not B.
No A are B.

The symbols A and B stand for categories. If one takes three such sentences where each pair contains one common category term, one can produce 256 different combinations. A couple of examples are

All A are B
All B are C
All A are C

and similarly

All A are B
No B are C
No A are C

It is easy to convince oneself that in both these examples, *if* the first two sentences are true, *then* the third sentence *must* also be true. It makes no difference which categories one chooses the symbols A, B, and C to stand for. This is a definition of *logically valid inference*. Aristotle analysed all 256 of the possible variations and showed that only 24 of them are such that, if the first two sentences were true, then the third must also be true.

Of course, syllogisms are only a small part of logic, but what is important here is that Aristotle realized that in analysing an argument one can pose two questions, namely (1) are the premises true, and (2) does the conclusion follow from the premises? In order for an argument to be deemed sound, both of these questions must be answered in the affirmative. Aristotle's priceless contribution is that he drew attention to the formal side of argumentation.

4. The Axiomatic Ideal The fourth important contribution to the development of the scientific approach is Euclid's axiomatization of mathematics. This was an enormous step forward from what had previously been known about mathematics. The students of the time before Euclid were taught a great deal about practical mathematics. That is to say, they were taught mathematical methods for solving practical problems associated with commerce, transportation, astronomy and measurement. In these situations, there can be raised two fundamental questions: (i) do these rules for calculation give accurate results, and (ii) in cases in which one is certain that the results are accurate, on what grounds do we base this mathematical certainty?

Euclid's contribution was to show that all the mathematics of the day (and a great deal more) could be logically deduced from a few axioms, i.e., sentences that were obviously true and did not require further justification. Mathematics became a *deductive* science: from secure premises (axioms), new knowledge was inferred using strict logical rules. That is to say, a mathematical proof gives certainty. Euclid was so successful with his axiomatic-deductive method that for a long time it was considered the archetype of how science should operate. A good example of Euclid's influence is Baruch Spinoza's magnum opus *Ethica Ordine Geometrico Demonstrata* (1675). As the title ('Ethics proven by Geometric Methods') suggests, Spinoza presents ethics as an axiomatic system, precisely as Euclid did with geometry.

Euclid's axiomatic mathematics led to a scientific ideal, *the axiomatic science ideal*, which can be characterized in the following way:

- Science aims to attain certain knowledge, not mere beliefs or opinions.
- Begin by setting up axioms, i.e. truths so obvious that they require no further justification.
- Next, deduce new truths from these axioms using strict logical methods.

In geometry and elementary arithmetic it seems possible to find axioms just by reflecting upon geometrical figures and numbers, but how is it done in e.g. mechanics or biology? In other words, how do we recognize true general propositions from which one could deduce empirical facts in e.g., physics or biology? Aristotle discussed this (*Posterior Analytics* 2.19.100a6-8); he argued that repeated perceptions of things and events resulted in true general propositions, such as 'man is an animal', or 'the natural state of a body is rest'. The process of generalising from repeated experience to such general propositions he called '*epagoge'*, usually translated as 'induction'. However, one should keep in mind that he did not view this activity as a kind of scientific inference or be represented as an inductive argument. (So the translation 'induction', given the usual meaning of this word, is somewhat misleading.) Rather he describes the process more like an automatic working of the mind in forming concepts and general propositions. Such general propositions could then be used as premises in scientific demonstrations. So all sciences have, in Aristotle's view, the same axiomatic-deductive structure, but the way we recognize axioms is different in different sciences.[4]

These four elements of our ancient Greek heritage – to seek explanations of natural phenomena in nature's internal properties, to argue on an even plane, to investigate the rules of argumentation and logical validity, and to build them into a logically consistent system–are all essential to the development of a scientific approach.

One may wonder why science did not flourish and develop is the days of the Roman Empire, or those of the Caliphate, since both of these cultures further developed the Greek culture. That this was not the case shows that the inheritance of these ideas, in and of itself, is not sufficient for a scientific revolution. However, Greek views were probably a necessary factor for the emergence of the scientific revolution almost 2000 years later. In order to understand just how monumental a change would ensue, it is perhaps prudent to look at how people thought during the middle ages.

1.2 The Medieval Worldview

The Roman Empire gradually declined and fell apart during the fifth century and the last emperor of Western Roman Empire, Romulus Augustus, lost power A.D. 476. The decline of the Western Empire was followed by a rapid decline of organised civilisation in Western Europe, and, astonishingly also a fast decline in population. (According to a quite modern hypothesis this decline in population was caused by a climate catastrophe, no summers for three consecutive years 536–538, which in turn might have been caused by volcanic activity from the Ilopango Caldera in central El Salvador, pouring out clouds of dust and hiding sunshine for several years.) The Eastern Roman Empire continued until A.D. 1453, and Plato's Academy in Athens still existed form some time, but the Emperor Justinian closed it A.D. 529. Thus, organised study of philosophical and scientific texts from antiquity almost disappeared both in western and eastern Europe. One may say that in Europe

[4] From a modern view-point, Aristotle's account needs two modifications; (i) From a series of observations one can generalise in many ways and the most natural one might be the wrong one, as is illustrated by his own mistaken theory of motion and (ii) mathematical axioms are nowadays not viewed as self-evident, but chosen by mathematicians for certain purposes.

the philosophical and scientific heritage from Greece was lost for several hundred years.

Population and culture slowly improved during the early Middle Ages and around 1100 A.D. higher education started to reappear in the Western world. The first institutions of higher education–the precursors of the eventual Medieval Universities – were medical and legal colleges in Italy. A century later, these colleges had expanded into universities with various faculties: theological, medical, juridical and *artium liberalium*, i.e., the liberal arts. Education in these universities consisted in studying canonical works from ancient times, like Euclid's *Elements* or Galen's works on anatomy. The view regarding knowledge at the time was that knowledge was obtained through the study of classical texts. That one could discover new information, or that the ancient thinkers could have made mistakes, were not viable options (Fig. 1.1). When Aristotle became known in Christendom around 1200 A.D. as a result translations from Arabic to Latin of Aristotle' works (the Greek heritage had been tended by Arabic scholars throughout the Dark Ages in western Europe), Thomas Aquinas (1225–1274) began what he saw as the important task of interpreting Aristotle in a way that would not conflict with Christianity. The result of Aquinas' toils was what we now call scholasticism.

A good, though quite late, example of the medieval view of how one accrued knowledge is given by Olaus Magnus' *Historia de Gentibus Septentrionalibus* ('History of the Nordic Peoples') (1555). Each and every chapter gives a fully detailed description of what past authors have to say on the matter and how they approached the subject. For example, Olaus Magnus often refers uncritically to what Tacitus and Jordanes wrote about Ultima Thule. One would think that Olaus Magnus, who was born and raised in Sweden, ought to have seen himself as a superior authority on Nordic history, geography and habits and manner of living than an author from antiquity who relied merely on hearsay. This lack of a critical stance illuminates the scholarly attitude of the time. Certainly one does find some passages where Magnus is describing his own observations, but nowhere does one find a passage of the form 'the famous author X's view on the subject is incorrect because I have investigated the matter and found it *not* to be the case'. Olaus Magnus is essentially a Medieval scholar who was unaffected by the new currents, which had at that time clearly begun to flow.

But one could find tendencies of experimental science already during the Middle Ages, especially in England, though these tendencies did not lead directly to any breakthroughs. This was perhaps because of a general distrust in the ability of a sinful mankind to obtain true knowledge of God's inscrutable creation. Or perhaps it was due to the general opinion that this sort of knowledge was relatively unimportant. One did not live long, hence that which was essential was the soul's salvation and whether one was destined for hell or heaven.

In studying the history of art, it is not difficult to see a gradual shift in attitude and the way of thinking throughout this period. Early in the Middle Ages, art is for the most part symbolic and often intended to elicit religious emotions. Yet there arises a tendency towards more naturalistic and representational art. A major step in this process was the discovery of the central perspective that allowed artists to

Fig. 1.1 An anatomy lecture given by the medical faculty at the University of Venice in 1493. The professor lectures from his desk, while two assistants dissect a body and point out the various organs. Neither the students nor the professor take part in the dissection, as the source of knowledge is the book from which the professor lectures

depict people and scenery more accurately. It gradually becomes clear that interest is shifting from art's religious function to the possibility of producing more accurate portrayals of the visible world. In due course, this shift leads eventually to the Renaissance. This major change in Western culture, which includes–among other things–a newfound interest in people and living conditions, helped lay the foundations of the scientific revolution.

1.3 The Scientific Revolution

An important element of the scientific revolution was a new approach to nature and the new form of knowledge thereof. People began to think that understanding natural phenomena was important and interesting, a view which few people, at least few among those who were literate and wrote, shared during the Middle Ages.

By the mid-1500s the belief that knowledge was to be found only in classical texts began to be replaced by the belief that one's own observations of natural

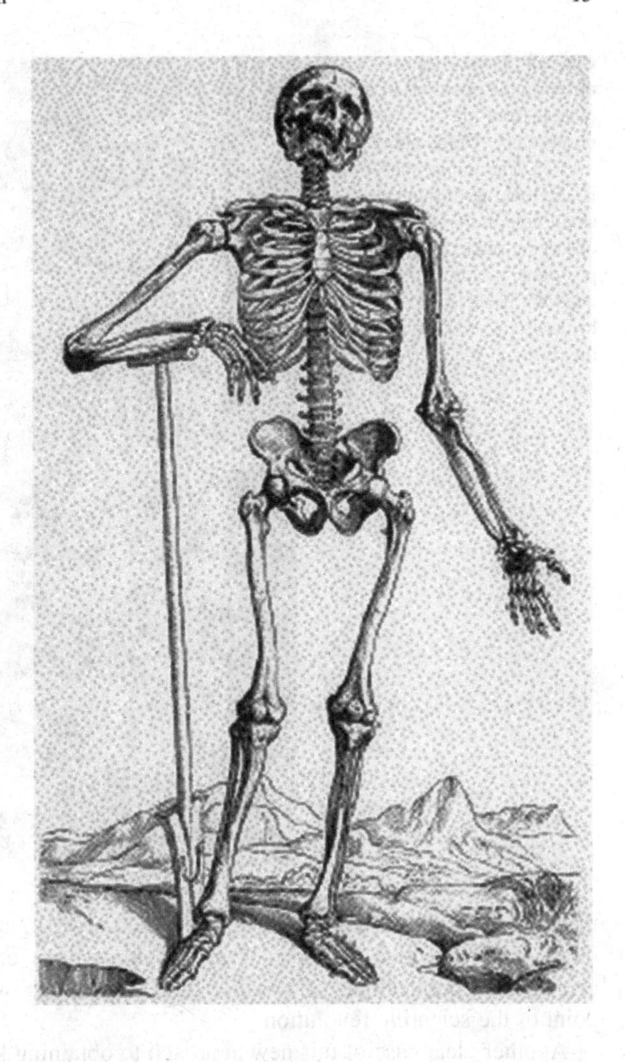

Fig. 1.2 An illustration from Vesalius' 'On the workings of the human body'

phenomena were necessary for knowledge of those phenomena. Closely related to this belief was the idea that the authors of antiquity could be mistaken, and that it was possible for contemporaneous scholars to correct those mistakes. The leading reformer for this cause was Andreas Vesalius, (1514–1564) of the University of Padua. His magnum opus *On the Workings of the Human Body* (1543) is generally considered the first modern text on anatomy. Vesalius' book was built upon comprehensive dissections, and he was able to refute many of Galen's views, which had been hitherto accepted as the authority on such matters. It marked the first appearance of a well-researched and essentially accurate description of the human body (Fig. 1.2).

In the same year, 1543, Copernicus published his new theory regarding the solar system. He was the first in modern times (Aristarchus of Samos, circa 200 B.C., had made similar claims, which inspired Copernicus) to state that it is the sun, and not

Fig. 1.3 Tycho Brahe observes the positions of the planets from his observatory, Uranienborg, on Ven. The image shows Brahe looking through a small hole in the wall, observing planetary positions via the large quadrant beside him, while his assistants record the time

the Earth, that is located at the centre of the universe, and that the Earth and the other planets revolve around it. One can argue quite confidently that these two events – the publications of Vesalius' and Copernicus' works – mark the starting point of the scientific revolution.

Another clear case of this new approach to obtaining knowledge is exemplified by Tycho Brahe (1546–1601). He devoted much effort to observing the positions and movements of celestial bodies from his observatory on the island of Ven in Denmark (now the island belongs to Sweden). Brahe's meticulous observational data was later used by Kepler in the development of his famous three laws of planetary motion (Fig. 1.3).

It is quite apparent that Brahe realized that he could obtain *new* knowledge, knowledge that could not be found in the classical texts. Why else would he have spent so many years making original observations? We should also keep in mind that the telescope had not yet been invented, and yet Brahe was able to pinpoint the position of planets and stars with an accuracy of 1 arc-minute, i.e. 1/60°.

It is apparent that mere observation is not sufficient for doing science, in that one has to somehow 'process' the observations (e.g. make classifications). The following passage from Bacon[5] is an example of this new perception of knowledge:

> Those who have treated of the sciences have been either empiricists or dogmatists. Empiricists, like ants, simply accumulate and use; Rationalists, like spiders, spin webs from themselves; the way of the bee is in between: it takes material from the flowers of the garden and the field; but it has the ability to convert and digest them. This is not unlike the true working of philosophy; which does not rely solely or mainly on mental power, and does not store the material provided by natural history and mechanical experiments in its memory untouched but altered and adapted in the intellect. Therefore much is to be hoped from a closer and more binding alliance (which has never yet been made) between these faculties (i.e. the experimental and the rational). (*The New Organon*, book 1, sec XCV)

Why did the scientific revolution begin in Western Europe at the end of the 1500s? Why did it not begin in China under the Tang dynasty, which was a cultural high period in China, or at the height of Arabic culture around the ninth century A. D.? Indeed, why did it not begin in the Aztec and Inca civilizations? In order to give a satisfactory answer to these questions one must, I think, apply both an *external* and an *internal* perspective. One must take into account both religious, economic and political factors (the external ones), as well as the ways that one obtains knowledge and the nature of knowledge in general (the internal ones). To properly merge these various aspects into a plausible explanation of why the scientific revolution began in Europe is an enormous project, and one that lies far beyond the scope of this book. However, one important question that we shall here discuss is whether there is a generally valid scientific method, and whether it was this method that began to be applied during the scientific revolution. My answer to this question is, reservedly, *yes*. There is a generally valid scientific method, the *hypothetico-deductive method*, of which more in Chap. 3, and it was this method that began to be applied in the work of Vesalius, Galilei, Harvey, Brahe, Kepler, Newton, and many more. The application of this method was an important factor, perhaps the most important, in the creation of the modern sense of scientific enterprise.

From the above discussion one might get the impression that the ancient Greek heritage had nothing to do with the development of modern science. This is, however, an incorrect conclusion. The scientific revolution consisted in a dramatic change in the medieval perception of knowledge, which was in large part inherited from the ancient Greeks. But, as we shall see in Chap. 3, there are ideas from antiquity actively operating at the emergence of the fully modern scientific method: the hypothetical-deductive method.

[5] Francis Bacon, (1561–1626), English philosopher and Lord Chancellor. He argued for empirical and inductive methods in science.

1.4 Theory of Science and Philosophy of Science

Theory of Science is a discipline that consists in investigations into the operations of science from both external and internal perspectives. However, in the remainder of this book we shall concentrate only on the internal perspective, which is the 'philosophical' perspective. We shall discuss such questions as: What is science? How does one distinguish science from pseudo-science? What is knowledge? What is a scientific explanation? What is a cause? What is a scientific hypothesis? What is a scientific theory? What is a scientifically acceptable observation? Hence the aim of this book is to provide the reader with a basic understanding of these issues, and the key concepts and views that philosophers and scientists have developed to answer the questions above.

The internal questions of Theory of Science belong to a part of philosophy; namely, philosophy of science. Like all philosophy, philosophy of science begins with wonder and bringing into question 'common knowledge'. One often speaks of science as something familiar, something that is easily identified. But, is this really the case? Does there exist a universally accepted definition of scientific practice? No, there does not; and even if it did, one could still ask how one might justify such a definition.

Theory of science, like the object of its study, also begins with questions regarding constitution and function. But Theory of Science differs from science proper in that there is no generally accepted school of thought from which to proceed. This fact can either be seen as a failure (Theory of Science has not been able to produce any lasting results) or as a distinctive character trait (Theory of Science *is* such that it does not produce solid foundations, rather it contributes to a critical review of science). In my opinion, the second alternative seems to be the most plausible.

Different questions within philosophy of science are relevant for different types of scientists. For example, the following questions are relevant for medical scientists: (1) What is a cause, in medical contexts? (2) What is the relation between the body and the mind? (3) What criteria should be used to distinguish between the practices of a charlatan and accepted scientific methods? The last question is a special case of the more general issue of distinguishing between science and pseudo-science. In discussing these three questions we are inevitably led to the questions regarding the nature of knowledge in general; what knowledge is, how it is that we obtain it, and how we can rely on this knowledge. Therefore, it is natural to proceed with a discussion of the nature of knowledge in general, and whether or not there is a common thread to be found in all activities we call science.

1.5 Summary

This chapter explains how our present conception of science is the result of a long evolution starting in ancient Greece. In particular four Greek ideas are pointed out as being crucial elements in scientific thinking: The first is the conviction that properties and events in nature could be understood as the result of natural processes, not as the effect of actions of gods and other supernatural agents. The second is view that knowledge could be achieved by rational argumentation, not from holy authorities. The third is the idea that a science should be systematic body of knowledge and the fourth component is the discovery of logic as a basic ingredient in rational thinking.

There was a clear empiricist trend in pre-Socratic Greek thinking, and to some extent also later. This empiricist outlook regained importance in western thinking around 1500 and was the central aspect of the scientific revolution.

Further Reading

Goodman, D. C., & Russell, C. A. (1991). *The rise of scientific Europe 1500–1800*. London: Hodder & Stoughton.

Grant, E. (1996). *The foundations of modern science in the Middle Ages: Their religious, institutional, and intellectual contexts*. Cambridge: Cambridge University Press.

Hannam, J. (2011). *The genesis of science: How the Christian Middle Ages launched the scientific revolution*. Washington, DC: Regnery Publishing.

Lindberg, D. (1992). *The beginnings of Western science*. Chicago: University of Chicago Press.

Russell, B. (1961). *History of Western philosophy*. London: Allen & Unwin.

Chapter 2
Knowledge

'Et ipsa scientia potestas est'
(Knowledge is power)
Bacon.
'Litterarum radices amaras, fructus dulces'
(The root of knowledge is bitter, but its fruit is sweet)
Cicero.

2.1 Introduction

Science aims to give us knowledge, scientific knowledge. According to the classical conception, the axiomatic ideal of science, scientific knowledge must be absolutely certain, the result of demonstration from general principles. But, alas, certainty has often proven to be an illusion; even Newton's mechanics, one of the most admired scientific theories and for a very long time believed to be certain, has proven not completely true. Should we still say that it is a piece of scientific knowledge? The answer depends, of course, on what conditions we should associate with the concept of knowledge. What is knowledge?

2.2 Knowing That, Knowing How and Acquaintance

The everyday use of the word 'know' denotes a variety of things. The first category consists of truths and we specify that kind of knowledge as 'knowing that', alternatively 'propositional knowledge'. Such knowledge is such that its content can be expressed by complete sentences. We have propositional knowledge if we *know that* something is the case. The content of this knowledge–that which we know–can be expressed in those complete sentences that follow 'that' in 'I know that . . .'. It can be obtained by listening to a lecture or reading a book, nothing more is needed.

L.-G. Johansson, *Philosophy of Science for Scientists*, Springer Undergraduate Texts in Philosophy, DOI 10.1007/978-3-319-26551-3_2

The second category consists of skills, such as riding a bike, playing a musical instrument, speaking a foreign language, etc., which we indicate by the term 'knowing how'. Such knowledge cannot be communicated only through language. This is sometimes called tacit or practical knowledge. For example, learning how to ride a bike requires both demonstration and practice. Much of our professional knowledge includes these two elements. Indeed, one would hope that, e.g., dentistry is ripe with both sorts of knowledge. Yet, it is often easy enough to draw a somewhat clear line between propositional and non-propositional knowledge, even within the same field of study. For example, most medical educations make a clear distinction between clinical and non-clinical courses, thus more or less reflecting the distinction between knowing that and knowing how.

The third category consists of knowledge of objects. In German, French, Swedish, Italian (and presumably other languages) the distinction between knowledge of objects and knowledge of truths is made clear by using different verbs. In German one distinguishes between 'wissen' and 'kennen', in French between 'savoir' and 'connaitre', in Swedish between 'veta' and 'känna' where the former words are used when we talk about knowledge of truths and the latter when we are concerned with knowledge of objects.

In common English there is no established corresponding distinction, which is a drawback for doing epistemology. I suggest using the word 'acquaintance' as term for knowledge of objects.[1]

The philosophical discussion about the nature of knowledge has almost exclusively been about knowledge of truths. The reason for this is that a correct analysis of this concept of is a highly controversial topic, whereas skills and knowledge of objects has not evoked much controversy. Indeed, when someone claims that he *knows that something is the case* (propositional knowledge), one is often want to doubt the reasons for this claim and have difficulties knowing whether it is in fact true or false. But if someone says that he *knows how something should be done* (a skill), or that he knows a certain person one often finds that the truth of such a claim is easy to verify. Thus, as we continue our discussion we shall concentrate on propositional knowledge, which we shall henceforth simply refer to as knowledge.

[1] Russell once introduced the distinction between knowledge by description and knowledge by acquaintance, which is not the same as that give here. According to Russell, the only things one had knowledge by acquaintance of were sense data. Talking about sense data was in my view a mistake; we have no knowledge about such things, since they don't exist. But ordinary physical objects, including persons exist, and these we may be acquainted with.

2.3 The Definition of Propositional Knowledge

The most common view among philosophers–derived from Plato–is that propositional knowledge is *justified true belief*. That is to say, if P stands for a particular proposition, and if X stands for a person who holds this proposition, then *X knows that P* if and only if:

- P is true,
- X can justify P,
- X believes P.

Each one of these criteria is a necessary condition for knowledge, and these three criteria are sufficient when taken together. Why is it that so many have stuck with this definition? An immediate answer to this question is that Plato's definition captures pretty well the everyday use of the word 'knowledge': the concept of someone knowing a proposition. Consider the following three dialogues.

First Dialogue

Charlie I know that New York is the capital of the U.S.A., I'm sure of it.
Lisa You're wrong, Washington D.C. is the capital of the U.S.A. It doesn't matter that you're sure of it, because that only means that you *strongly believe* in your claim, and that's not the same as *knowing*. In order to *know* something, that something has to be *true*.

Second Dialogue

Charlie I know that it's going to rain tomorrow.
Lisa How do you know that?
Charlie I just know it!
Lisa If you can't explain *how* you know that it's going to rain, then you don't know it. Even if it is true that it will rain tomorrow, you don't know that unless you can explain how you came to your conclusion, or belief.

Third Dialogue

Charlie I know that the speed of light is constant and independent of the speed of the light source, but I don't believe it.
Lisa Why don't you believe it?
Charlie Because common sense tells me that the speed of a light ray should be the sum of both the speed of the light source and the speed of the light ray when it is emitted.
Lisa Then why do you say that you know that the speed of light is constant?
Charlie Because I know that Einstein claimed that it is the case, and all physicists believe him.

Lisa But if you don't believe it yourself, then you can't claim that you *know* that the speed of light is constant. All you can claim to know is that *Einstein and other physicists claim that the speed of light is constant*, which is completely different.

These dialogues are perhaps a bit artificial, yet they seem to reflect how we typically view the proper use of the phrase 'I know that...'. If they are accepted as accurate reflections of every day speech, then they show that each of the criteria above are necessary conditions for knowledge. Some philosophers have constructed situations that purport to show that together the above criteria are not sufficient for knowledge; this is the famous Gettier's problem, which I'll discuss in Sect. 2.4. But first we shall discuss some implications of these critieria.

2.3.1 *P Is True*

If one knows that something is the case, one may make a claim expressing this fact and that claim is then true. Another way of stating the same thing is to say that when one knows that something is the case, one is simply describing a part of the world as it actually is; since, to say that a proposition P is true is equivalent to asserting P.

This relationship between knowledge and truth is essential for our actions, when we plan what to do; for if knowledge was independent of truth, what one knows would be of no use in guiding one's actions. This is a fatal consequence for all situations in which one needs to *use* knowledge for some purpose. An example that illustrates this point is the difference between the actions of the American Federal Reserve in 1929 and 1987. During the stock market crash of 1929, the Federal Reserve kept interest rates high (to defend the gold standard) and let a large number of banks go bankrupt. This resulted in people loosing faith in their banks. This loss of faith prompted people to remove their savings; thereby, making it impossible for the banks to lend out money or make investments. Ultimately, all this culminated in the great depression and a 60 % decrease in the GDP of the USA over 3 years. Not until 1941 was the GDP back to the same level as 1929. This was certainly not intended and it is clear that Federal Reserve did not know enough about the mechanisms of the economy.

In 1987, with much better insight as to the workings of the economic system, people realized that the role of the Federal Reserve was to ensure that the banking system functioned properly, and to make sure that people believed that their money was secure. The Federal Reserve took actions to ensure this security, and achieved the desired results. For even though the stock market had a sharp decline for 1 week in 1987, no depression resulted and the crisis was over within a year.

But was the knowledge about the economic mechanisms good enough? There are discussions about the decision to let the Lehman Brothers bank go bankrupt in 2008, since it started a long period of low growth. Some say that the decision makers had not properly learnt the lesson from the great depression. I have no considered opinion about these matters; it only further highlights the point that true justified beliefs, i.e. knowledge, is crucial for proper action.

Sometimes people equate 'true' with 'true for me'. This is a mistake. Truth should be understood in an absolute and objective sense: a proposition is true, or false, *simpliciter*. The truth of a proposition is completely independent of the beliefs and cultural context of the person who utters it. Indeed, one must clearly distinguish between 'P is true' and 'X believes that P is true', for it is obvious that X can believe P, i.e., that P is true, when in fact P is false, and vice versa. This implies that a person can believe that he knows some proposition, even though the proposition is false. Indeed, it often happens that long after one has claimed that a proposition is true, one realizes that it is false; at which point one should admit the mistake and withdraw the claim to know.

When talking about older theories some people are inclined to say that such theories was true, but now they are false. I have, for example heard many students saying that Ptolemy's geocentric theory about the world, put forward in his *Almagest*, was true in his times, but now it is false since it contradicts the heliocentric theory which is true. Ptolemy's theory is based on observations of the motion of the sun, the moon, the planets and the stars. They all appear, from the view-point of an observer at earth, to move around the Earth in circular motions.[2] With Ptolemy's theory, one could make fairly accurate predictions of the positions and motions of all celestial bodies visible with the naked eye, so Ptolemy's theory was supported by observations and had predictive power. Indeed, prior to the invention of optical instruments, it was not possible to make any observation that would contradict his theory. Almost anyone alive at a time prior to the invention of the telescope (early 1600s), would easily be convinced that one could obtain knowledge of the universe by studying Ptolemy's book. However, we now know that Ptolemy's theory is false, and that all those who claimed to have knowledge of the universe through the study of Ptolemy's work were mistaken.

We may say about ancient astronomers that they had good evidence for their theory and they believed it, but we should not say that they knew how the world system moved. For if I would say that they knew, I would accept that their belief is true, which I do not.

Some are inclined to say that Ptolemy had knowledge of the planetary system, but that we *now* have different (or better) knowledge. This statement implies that one accepts that Ptolemy had *false* knowledge. But false knowledge is not knowledge, just as counterfeit money is no real money (Fig. 2.1).

A common argument against an objective, absolute understanding of truth builds on the fact that humans have developed many different cultures and conceptual frameworks. Some claim that there is no such thing as absolute truth; that all truths are relative to one's culture, paradigm, theory or research domain. The argument is

[2] This is a simplified picture; Ptolemy was forced to add so called epicycles, circles on the circles, in order to get empirically adequate description of some of the planet's motion.

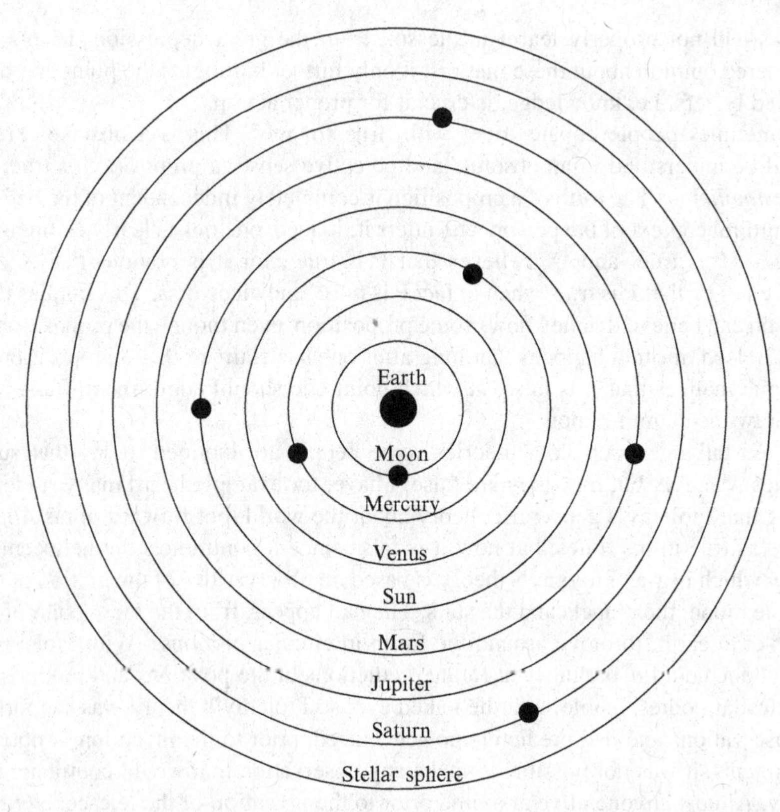

Fig. 2.1 The basic structure of Ptolemy's geocentric worldview (No epicycles are included in this figure, see Fig. 3.2)

that we cannot say anything about the way the world is without the use of language and our languages are human creations. From this premise one arrives at the conclusion that truth is language dependent. Another way of stating the argument is that humans live is different conceptual frameworks; and so there is no such thing as truth in and of itself, but only truth in some conceptual framework.

There are plenty of arguments against this line of reasoning, of which the first was already laid forth in Plato's *Theaetetus*. This dialogue contains a discussion of Protagoras' claim that *things are as they appear to be*. This claim is a version of the relativistic thesis that there are no objective truths. Theaetetus, a young man who finds himself in conversation with Socrates, puts forth Protagoras' claim as a possible definition of knowledge. Socrates then asks Theaetetus whether everyone is in agreement with Protagoras regarding this matter, and Theaetetus answers that most people are not. Socrates then responds that Protagoras must admit that his claim is incorrect, for it says that things are as they appear and since to Protagoras' opponents the view appears false, it is false. This leads one to realize that the thesis that *all* truths are relative is self-contradictory. We cannot consistently say that *all* truths are relative.

Another interesting argument against relativism, given by Donald Davidson,[3] is based on observations of human language-users. It is presented here in abbreviated form:

1. To claim that other humans live in other conceptual frameworks (and that truth is relative to some conceptual framework) than our own implies that we claim that they think and speak.
2. To correctly claim that people think and speak requires that we know they are actually saying something, and not just producing sounds.
3. In order to know that people are saying something, we have to at least know something of what they mean.
4. In order to know what people mean, we must be able to translate some of their utterances into our language.
5. In order to translate their utterances into our language, we must ascribe to them a set of beliefs, desires, attitudes, and a way of connecting these mental elements with each other.
6. In order to ascribe to people such mental elements, we must first assume that they share with us some background beliefs, desires and principles of thought.
7. But to have in common such a reference frame (beliefs, desires, and a capacity for knowledge) is to be situated in the same conceptual framework.

A key step in this argument is premise 5. The basic idea is that interpretation of what other people are saying requires that the interpreter ascribe to those people a large number of true beliefs about reality. (This is an instance of the *principle of charity*, a core element in Davidson's theory of interpretation.) In addition, the interpreter must also assume that the people, whose utterances are to be interpreted, are rational: otherwise one cannot make any sense of the content of those utterances. If this premise is accepted, then it seems that the conclusion is sure to follow.

The conclusion of this argument is that in order to claim that other people 'live' in different conceptual worlds, or frameworks, we must be in agreement, with respect to these people, about a considerable number of propositions, regardless of one's conceptual framework. Thus both Davidson's and Plato's arguments show that *all-embracing* relativism is not feasible.

There are thus strong arguments for the claim that the truth of a proposition depends upon how the world is actually constituted and not on how we perceive it. Aristotle formulates this idea as a definition of truth in his *Metaphysics*, (book 4, part 7):

> To say of what is, that it is not, and to say of what is not, that it is, is false, while to say of what is, that it is, and to say of what is not, that it is not, is true.

This passage is often interpreted as an expression of a *correspondence idea*: if a sentence is true, then there is a correspondence between the sentence and what the

[3] Donald Davidson (1917–2003), American philosopher of language and psychology. The argument is to be found in his paper 'On the Very Idea of a Conceptual Scheme', printed in his (1984).

sentence is about. It is quite difficult–some would say impossible–to describe more precisely the nature of this correspondence, and such an inquiry is beyond the scope of this book. Instead I shall move on to a discussion of the various consequences of this idea.

One such consequence of the correspondence theory of truth is that a proposition can be false even though we may have very good reasons for believing that it is true (and vice versa). This consequence opens the door for the sceptics, who want to argue that we can never be certain that we know anything. If the sceptics are correct, then the search for truth becomes pointless, since one can never be certain that it is ever attained. This seems to be a very strange conclusion. Indeed, consider humanity's canon of reasonable beliefs; there are good reasons for believing that a great deal of these are knowledge by virtue of being true. Yet, there are also good reasons for believing that some part of this canon is false. In other words, some (hopefully small) part of what we call knowledge is not knowledge. Of course, we do not know which parts of what we call knowledge are false: were we aware of such falsities, we would certainly separate it out and label it accordingly. (This very activity is often a result of research; one discovers that something previously accepted as knowledge is false). However, it would hardly be rational to throw away everything we take to be knowledge on account of a few bad eggs hidden among the good. The rational stance is surely an attitude of provisional belief: every proposition we call knowledge is probably true, but there is a risk of it actually being false, and so we should be mentally prepared to change our beliefs if new evidence presents itself.

A sceptic would ask: how do you know that most of what you call knowledge is true? Maybe it is all false! My answer to this question is simply that, in practical situations, even the sceptic takes an enormous amount of his beliefs as unproblematically true. I shall show this by giving a few examples.

Suppose the sceptic has a TV, and suppose that he pays for cable programming. The sceptic, in such a practical situation, believes, as a matter of fact, that he will continue to receive TV programming several days into the future. Similarly, when the sceptic purchases a carton of milk from a supermarket he believes that there is milk in the carton, and he would be quite surprised to find that it contained anything else. The sceptic drives to work with the belief that the building in which he works still exists, and that he was not fired overnight. Upon reflection of these examples it seems that in order doubt something, one must take a great many things for granted. One can only doubt a proposition with respect to some background knowledge that one takes for granted, which implies that total scepticism is an incoherent view.

A more modest sceptic would perhaps say that he does not doubt that some proposition is true, but that he merely refrains from believing any propositions. However, in practical situations, as we have seen, even the modest sceptic would act as if he does believe some propositions to be true. Thus even the most compromising of sceptics behaves as if he believes, in all practicality, a great number of propositions. Since there are very good reasons to hold that both habitual and deliberate actions are based on beliefs, independently of their expression, there are very good reasons to believe that the sceptic believes a great many propositions, despite his assurances to the contrary.

In summary, we attribute to ourselves a great deal of knowledge; albeit, some of which, unbeknownst to us, is not actually knowledge by virtue of being false. This is the basis of the *provisional nature of knowledge*: since we do not know which of those propositions we call knowledge is false, we must be prepared to rethink everything, even though we have good reasons to believe that most of what we take ourselves to know is true.

2.3.2 Good Reasons for P

What constitutes good reasons for a proposition P depends on the kind of proposition P is. We can broadly distinguish three different types of propositions that are justified in completely different ways: (i) *mathematical*, (ii) *observational*, and (iii) *theoretical*. The line between observational and theoretical propositions is a controversial issue that shall be left for later (The dependence of observational propositions on theory will be discussed in Sect. 4.4).

(i) To give good reasons for believing a *mathematical proposition* is to give a proof. If a mathematical proposition has not been proven, then there is not sufficient reason to believe it is true.

(ii) It is less clear what constitutes good reason to believe observational propositions. In everyday situations, the fact that the person who is making the claim saw the relevant event take place is often considered a good reason for us believing him. However, anyone can make a mistake when it comes to one's observations, and we often delude ourselves. Richard Wiseman discusses these things in his *Paranormality*.

In a court of law a single direct witness of a person committing a crime will often not suffice to prove a case. An interesting example is the Court of Appeal's treatment of Lisbet Palme's designation of Christer Pettersson as her husband's killer. The prime minister of Sweden, Olof Palme, had been shot by a gunman at the centre of Stockholm 1986, and Christer Pettersson was tried for the murder and convicted in the first instance. But the circumstances of her identification of this person were such that there was reasonable doubt; and therefore Pettersson was acquitted by the court of appeal on account of lack of evidence.

However, if several unconnected people had observed the shooting, and all these people were of sound mind and in circumstances favourable for reliable observation of this kind of incident, then their collective testimony would have been enough to count as good reason for believing Pettersson guilty of the crime.

Whereas a single person's testimony often fails to suffice in a court of law, often the testimony of a single scientist is taken to suffice as good reason to believe some result in the sciences. For example, one scientist's observation of a measurement on an instrument is sometimes considered a good reason for believing that the observational proposition is true. But, of course, we cannot completely dismiss the possibility of fraud or mistake.

(iii) What constitutes a good reason for believing a theoretical proposition is far more difficult to say. Indeed, one could say without too much fear of contradiction by one's peers that the most discussed question in the philosophy of science regards precisely this issue: under which conditions is a scientific theory sufficiently strong for commanding general confidence? This question will be the central topic of the next chapter.

This criterion for knowledge–good reasons for believing a proposition is true–is perhaps the least controversial as a general condition. If a person believes some proposition P, which happens to be true, but cannot give good reasons for his confidence in the truth of P, then we would hardly say that this person knows P. Instead we would perhaps call it 'hearsay', or that 'he is just repeating what he heard'.

2.3.3 What Does It Mean to Believe That P?

To believe that a proposition is true is the mental aspect of the concept of knowledge. Notice that belief is not essentially religious. All people believe a large number of different things about the weather, the future and themselves. Belief is a mental state, similar to hope, doubt, etc.: i.e., it is a *propositional attitude*. (See Chap. 12).

It would be completely unnatural to say of a person that he knows that P, yet does not believe that P. In particular, if we accept the rest of Plato's definition of knowledge, this would amount to our claiming that he has good reason to believe P, yet he does not believe P. How could this be rational? (The philosopher G. E. Moore once remarked that it is paradoxical to say: It's raining, but I don't believe it.)

Note also that this criterion implies that if we want to say that an animal, or machine, has (propositional) knowledge, we must ascribe to that animal, or machine, beliefs concerning, for example, its immediate surroundings. There are various views on this subject: some think that there are no principal differences between humans and animals in that at least *some* animals are capable of belief, whereas others think that *only* humans are capable of belief. Additionally, even if one ascribes beliefs to animals, it does not follow that animals can have propositional knowledge, i.e., knowledge that is transmitted through language. The most reasonable view is, it seems, that animals only have non-propositional knowledge (skills). Even if we can 'speak' to dogs and various other animals, it is doubtful that what we teach them by 'speaking' to them is the kind of knowledge that we humans acquire when we read a book or listen to a lecture.

Can computers have knowledge? Can they have beliefs? In today's world it does not seem reasonable to say that computers could have these capabilities. What would we say about a futuristic computer that is capable of anthropomorphic behaviour? Would we still say that there is a principal difference between humans and computers? This is a controversial question within AI (Artificial Intelligence) research as well as within the philosophy of mind, and there is little consensus among the various researchers and philosophers. The only thing that can be said, in general, is that if we want to retain Plato's definition of knowledge whilst allowing for future machines having knowledge, then we would also have to allow for future machines having 'beliefs'. On the other hand, if we think that computers are in principle incapable of having beliefs, and if we want to retain Plato's definition of knowledge, then we must hold that computers are likewise incapable of knowledge, irrespective of their performance.

For many it is obvious that computers cannot believe, have knowledge or think; and that these abilities are uniquely human. However, we should note that whenever we say that a person believes something, knows something, or is intelligent, that we do so on the basis of her apparent behaviour. We cannot, after all, observe a person's mental processes. If we were to use the same criteria (in terms of observed behaviour) when considering whether a computer can have knowledge as we do when considering whether a human can have knowledge, then it seems that computers could have knowledge, at least in principle. The English logician Alan Turing (1912–1954) formulated this idea as the Turing test, which is a criterion for intelligent, human behaviour. The Turing test is carried out in the following way: suppose you type questions on a keyboard that is in some unknown way (for you) connected to other devices. The printer then prints out the answers to your questions. You are allowed to ask whatever you want and as many questions as you please. If you are unable to tell whether it is a human or a machine that is answering the questions, and it is a machine that is answering, then that machine is exhibiting human mental capacities. That is to say, the machine in question has knowledge and it can think.

This test falls short of any generally accepted criteria for thought. But those who reject it are faced with two options: either they must construct some other criteria for thought, or else maintain that rational thought is a uniquely human capacity. If one takes the latter view, then, of course, every such test to see if a machine can think is completely misguided.

In summary, we have discussed one of the most widely accepted definitions of knowledge. Application of this definition has brought with it various consequences as regards our parlance. The point of this discussion is partly to promote a well thought out and consistent use of the concept of (propositional) knowledge, and partly to clarify the relationship between knowledge, truth, justification and belief.

2.4 Can One Know Without Knowing that One Knows?

A common argument against the requirement that knowledge must be true is the following: since you admit that in many situations you cannot be sure that a proposition P is true, even if P is true, then you do not know that you know P; and thus you do not know P. This is a logical mistake. The argument has the following form:

1. If X knows that P, then X knows that X knows that P.
2. It is not the case that X knows that X knows that P.

Therefore,
It is not the case that X knows that P.

The mistake is in the second premise. The content of the claim that one knows that one knows that P is that

- One believes that one knows that P,
- One has good reasons for saying that one knows that P and
- It is true that one knows that P.

The first criterion is plausibly met; if a certain person, X, knows P, then P is a justified, true belief of X. It follows that if one attributes to X minimal capacities of self-reflection, then it is reasonable to suppose that X believes she know P. Similarly, one can say that she has good reasons for saying that she knows P. But is it true that X knows P? Yes, if P is true, X believes P and is justified in believing P, then X knows P. But this is the same as saying that it is true that X knows P. Thus one can say that one knows that one knows P, even though one has no sure method to *determine* whether P is true. It is, *in fact*, enough that P is true.

In addition, it is a controversial question whether premise 1 is a valid principle; many claim that it is a logical truth, (an axiom in epistemic logic) while others doubt this claim. The critics do not want to doubt that in individual cases it can be correct to say that if one knows P, then one knows that one knows P, but they do doubt that the principle holds for all cases. The dispute is about how to formalize the logical principles of the concept of knowing. Those who doubt that premise 1 is a valid principle have another argument against the claim that one does not know a proposition when one lacks a guarantee of the propositions truth.

2.5 Reliabilism

The conception of knowledge as justified true belief has been questioned the recent decades. The seminal paper in this development was a short paper by Edmund Gettier, 'Is knowledge justified true belief?' (1963). He constructed a couple of examples where the three conditions for knowledge are clearly fulfilled, but yet we still would be unwilling to say that the proposition under discussion was known by the speaker. This suggests that the three conditions are not sufficient and that some amendment is needed. But, alas, all efforts to add something have failed due to the additional clause excluding whole categories of propositions that we would recognize as clear examples of knowledge.

It has become increasingly clear that the problem is not truth, not belief, but justification: more precisely, justification of empirical propositions. This has inspired an increasing number of epistemologists to replace justification with a more descriptive condition, viz., that the beliefs called 'knowledge' have been arrived at using a reliable procedure; thus the label 'reliabilism'. The general idea is that whether a person can be said to know a proposition P or not, does not depend on him being able to produce good reasons, but whether he *in fact* used a method for arriving at his belief that has proven reliable. Which methods should qualify as reliable is currently under debate, but the tendency is clear: a lot of people are adopting some variant of reliabilism.

This trend is but one aspect of the naturalistic tendency in current epistemology, viz., a rejection of the notion that epistemology can be an independent basis for the other sciences. Naturalists view epistemology as an integral part of science, no more basic than cognitive psychology for example. The interested reader may consult *Stanford Encyclopaedia of Philosophy/knowledge – analysis* (http://plato.stanford.edu).

Some authors e.g. Luciano Floridi (2011) claim that the Gettier problem cannot, as a matter of principle, be solved. Floridi's conclusion is to focus attention on the concept of information instead; this is more basic than the concept of knowledge.

2.6 Data, Information, Knowledge

It seems that scientific knowledge in many areas, e.g., in palaeontology, genetics and medicine, is increasing with an ever-increasing pace. New technologies in e.g. DNA sequencing and new possibilities to store, retrieve and analyse enormous amounts of data has made it possible to gain knowledge in areas previously believed to be inaccessible.

Using modern technology it is possible to collect data, transform data sets to information and then arrive at more knowledge. Such processes have always been important in science, but with the rapid development of computers and computer technology it has come to centre stage in some sciences. Therefore the relations between the three concepts, *data, information* and *knowledge*, deserve some scrutiny.

Some philosophers, the first being Kant (as far as I know) observed that the concept of information is somehow more basic than that of knowledge. In modern times Gareth Evans made the same observation, according to Michael Dummett:

> Evans had the idea that there is a much cruder and more fundamental concept than that of knowledge on which philosophers have concentrated so much, namely the concept of information. Information is conveyed by perception, and retained by memory, though

> also transmitted by means of language. One needs to concentrate on that concept before one approaches that of knowledge, in the proper sense. Information is acquired, for example, without one's necessarily having a grasp of the proposition which embodies it; the flow of information operates at a much more basic level than the acquisition and transformation of knowledge. I think that the conception deserves to be explored. It's not one that ever occurred to me before I read Evans, but it is probably fruitful. That also distinguishes this work very sharply from traditional epistemology. (Dummett 1993, p. 186)

But what more precisely is the relation between information and knowledge?

The first thing to notice is that the word 'information' is used in, at least, three distinct senses. The first sense is *semantic information*. We use the 'information' in this sense in the following examples:

- 'The study counsellor informed the student about the job prospects for lawyers.'
- 'The allied forces during WWII collected information about German plans by decoding their encrypted messages.'
- 'By studying the map carefully we got information about the best way to drive.'

In these examples the information talked about can be expressed as complete meaningful sentences, which is why it is called semantic information.

It has been some debate concerning the veracity of semantic information; must it be true or not? It is now a general consensus that false information is no real information. False information can be compared with counterfeit, i.e. fake money. Fake money is no real money and similarly false information is no information. The reason is that in neither case can it (false information or counterfeit) be used for what it is supposed to be used for. Money is used for buying things, and false money cannot be used for buying anything, it is worthless. (If you succeed in getting an item by using counterfeit, you did not really pay; you deluded the seller, and hence you did not buy the item.) Information is used for making plans and decisions in order to achieve our goals. But false information is useless for this, or worse.

In both cases we may be unaware of the real state of affairs. We may mistakenly believe that some piece of information is true and make decisions based on this. The risk of failure in such a case is much higher than if we had true information. Similarly, we may be unaware that a note in our possession is counterfeit and we try to use it for a purchase. Clearly, the risk of failure in our endeavour is higher than if it was real money; we are even in danger of being accused of being aware about the counterfeit and prosecuted.

It may be observed that semantic information need not be linguistic items. A map contains geographic information using a lot of symbols, lines, dots, curves, etc. By interpreting these symbols we can formulate sentences that express information contained in the map.

The second sense of information is usually called 'Shannon Information' after the seminal work of Shannon and Weaver, *The Mathematical Theory of Information* (1949). In this book they conceived information as a quantity. They studied the question how much information that can be transmitted from a sender through a cable to a receiver without distortion. In their resulting communication theory they defined information in terms of probability. The basic idea is that the amount of information in a message is higher the less probable the sequence of signs is. Information in this sense may be called negentropy, i.e. the opposite to entropy.

Shannon information is the intended sense in the following sentences:

- 'My memory stick can store 16 GB information',
- 'The flow of information operates at a much more basic level than the acquisition and transformation of knowledge.' (Dummett quote above)

The third sense of 'information' is somewhat similar to semantic information, but where information is said to be contained or stored in a system without any higher cognitive capacities. Consider these examples:

- Our DNA contains information about hereditary traits.
- Migrating birds spending the winter in Africa obtain information about the time for flying to northern countries from the length of the day.
- My computer got information that a new version of the virus protection programme was available.

In such cases there is no mind involved. We might say that in these cases it is systems with feedback mechanisms that under certain inputs perform a certain output, and we call the input 'information' because it can be described in functional terms. We would not say that these systems obtain knowledge about the conditions; but we find it natural to say that they obtain information and perform certain operations as a result of the information.

But why call the input 'information'? The reason seems to be that we can describe the input as being *about* something, often the state of the environment. It has content. Or rather, when *we humans* describe the input and the workings of the system we find it natural to talk as if the information-containing system consciously sent messages to us humans; we say that the systems obtain information, transmit information or store information about something, as if it were like a human mind. The core feature of this use of the word 'information' is thus its *aboutness*, its intentionality. This notion will be further discussed in Chap. 5.

Finally *data*. It is common in computer science to say that information is data with meaning. This is ok as far as it goes, but what is 'meaning'? And how do data acquire meaning? It seems that minimally it means that meaningful data becomes information when we have been able to formulate declarative sentences expressing the information that is obtained from a data set.

Almost anything can be data. In order to obtain data from e.g. a story, from light from distant stars, or from the result of an experiment, we need to divide the stream of sounds, lights, or states of detectors into distinct items. When using written text as data source one must divide the string of linguistic signs into distinct items, such as words or longer or shorter expressions. Radiation from universe may be divided according to position (using a coordinate system for identifying position in the sky) and frequency distribution for example. Detector states, clearly distinct from each other, must be defined before any meaningful measurement can be done. In short, in order to obtain a data set, we need to define a principle for dividing up something into distinct pieces. Hence from a conceptual point of view, discerning data and collecting a data set presupposes that we have a prior principle of making distinctions within a phenomenon. And making distinctions is always driven by presuppositions about what might be relevant and not for the research question. Sometimes we have lot of background knowledge from start, for example when studying the cosmic background radiation it seems obvious to sort data according to intensity and frequency in different directions. But in other cases it is far from obvious: listening or reading a story in order to collect empirical data there is no obvious way of transforming this story to a data set. Should we count frequencies of certain expressions? Should we look for grammatical forms? Should we study the order of types of expressions? It depends on purpose, of course.

The first sense of 'information' is that which is most close to what we mean by 'knowledge'. Knowledge and semantic information can be expressed as true statements being about something. One could say that semantic information is the proper concept when we talk about the content of knowledge states of humans, but disregard that these knowledge states are mind states or whether the subject can produce good reasons for the content of his belief. One could say that semantic information is that which is expressed by the sentence p when we say of a person that she knows that p. In short, a piece of knowledge is a piece of information for which the knower can provide good reasons.

2.7 The Philosopher's Versus the Sociologist's Concept of Knowledge

The discussion has hitherto been about what one *ought* to call knowledge. Claiming that the *correct* meaning of the expression 'X knows that P' is that P is a true, justified belief of X is stating a norm. However, one can take a totally different perspective by applying a more sociologically oriented approach. Accordingly, knowledge is seen as merely that which humans within a particular society *call* knowledge. This means that one places 'true' and 'considered true' on the same footing. If one were to change the societal perspective, then what was knowledge may no longer be considered knowledge. There are two immediate interpretations of this sociological perspective. The first is that it is merely another version of relativism, in the sense that what a culture holds to be knowledge *is* knowledge, and there is no way of distinguishing between *real* knowledge and what is *taken* to be knowledge. The second interpretation is that sociologists are operating with another concept of knowledge, 'what a certain culture considers to be knowledge', and that they do not take a stance as to whether it is actually knowledge or not, or even whether it is possible to make a distinction between knowledge and what is taken to be knowledge. It may seem a legitimate viewpoint for a sociologist to be agnostic when it comes to this question, since sociology is not philosophy. However, if this agnosticism is universalized, then it becomes applicable to those propositions the sociologists themselves produce, and we are led to ask whether the sociologist's claims are supposed to be true regardless of culture, or are they merely opinions within his research group, discipline, university or society? Suppose that the sociologist answers that the latter alternative is the case, which he ought to do, since the first alternative leads to inconsistency in the sociologist's view. Then we ask ourselves, what is the point of his research? Does he want to solicit his own cultural domain? It is obvious that total relativism eventually leads to a complete surrender of any distinction between scientific argumentation and propaganda; or, for that matter, between science and religious dogma. But some restricted form of relativism, where some inquiries are viewed as resulting in objective knowledge, whereas others are held to express mere opinions, is possible.

Helen Longino has in her (2002) discussed these matters. She argues that the dichotomy between the rational and the social aspects of knowledge depends on some presuppositions that can be refuted.

A number of sociologists, for example Bruno Latour, have introduced the norm that sociologists ought to study societies and groups without taking a stance as to whether the claims of those societies are true or not: one should be impartial. This may seem to be a sympathetic view; that one should not act as a know-it-all and dictate as to who is right or wrong. If one agrees with this norm, then it follows that one cannot, in principle, distinguish between what is considered knowledge in the society under study and what actually is knowledge. This implies that one cannot appeal to factual circumstances when one wants to explain something within the society or culture being studied. Among other things, this means that one cannot explain the results of various decisions or behaviour by saying, for example, that it was based on a misunderstanding, since the idea of a misunderstanding presupposes the distinction between truth and falsity. This is, in all practicality, an unreasonable position.

2.8 The Expression 'It Is True for Me that...'

One sometimes hears people say that something is 'true for me', or 'it is true for her'. Obviously one wants to say something more than that he or she believes the proposition in question, otherwise, one would simply say that 'I believe that...', or 'I believe that...is true'. When one says that '...is true', it is because one wants to claim something stronger; namely, that one's claim is not just a private opinion, but also that it is objectively true. So what is being said when one adds 'for me'? One gets the impression that the speaker initially wants to claim something more than a private opinion, but then later signals a retraction when he adds 'for me'. Thus, uttering the expression 'it is true for me' seems to be a pragmatic inconsistency. One is both making and taking back the claim that the utterance is objectively true.

There are, however, situations where the expression 'it is true for me' can be given a reasonable interpretation. I am thinking of a case where one talks about how one experiences a situation and wants to assert priority of one's own experience. An example is the discussion that followed the publication of Jan Myrdal's[4] autobiographical books *Childhood* and *Twelve Going on Thirteen*. In these books, Myrdal, a well-known Swedish author with well-known parents, presents his mother in a very unfavourable light. Myrdal describes himself as a boy longing for his mother's love and attention, which he hardly ever gets. Since his books came out, Myrdal's sisters have also published autobiographies in which Myrdal is presented as a terribly egoistic boy, who is always jealously manoeuvring to get his mother's undivided attention. In response to these claims Myrdal answered, 'It is my truth that I describe in my books.'

This answer is impossible to criticize. The only thing one can say is that it would have shown a sign of impartiality and self-awareness if this phrasing had been more clearly presented in his books. For example, Myrdal could have written, 'I *experienced* my mother as being false', instead of the infamous 'she was as false as a three-crown coin'. However, Myrdal can defend himself in that, in his books, he pointed out that it is his personal perspective that is being presented.

This example shows that, when the question regards how one should describe a social situation, one cannot ignore how the situation is experienced by those involved. One could, perhaps, defend Myrdal's parents with the argument that they did not in any obvious way deviate from the accepted norms of childcare of the time and culture, but this does not suffice. Suppose that Myrdal, with respect to both past and present day norms, was a horribly self-absorbed person, and that his parents engaged with him more than usual at that time. Would this imply that we could claim that he was wrong to portray his mother as such a mean and uncaring person in the description of his childhood experience? No, hardly. Questions regarding how an event is experienced and how some event actually unfolded are, in this case, the same question. The objective reality we want to describe is precisely how different people *perceive* their relations to one another.

There is reason here to make a distinction between two different senses of the objective-subjective contrast. The first sense is *epistemological*. It is that a claim is objectively true, or objectively false, if the truth or falsity of that claim does not depend on who makes the claim. Another sense of the objective-subjective contrast is *ontological*. A phenomenon, circumstance, event or object, in short, anything we can talk about, is objective in the ontological sense if its existence is independent of any individual's mental states. Otherwise, it is ontologically subjective. A person's attitudes, feelings and thoughts are subjective in the ontological sense. It follows that one can make objectively true or false claims about these phenomena, i.e., about subjective experiences.

What is the criterion for truth when we ask, 'what was the relationship between Jan Myrdal and his mother?' The only reasonable answer is to say that the truth about the relationship between mother and son, in this case, has to do with precisely

[4] Jan Myrdal, born 1927, is in Sweden a well-known author and son of Gunnar and Alva Myrdal, both being well-known to the public. Gunnar Myrdal was awarded the nobel prize in economics 1974 and Alva Myrdal was active in the peace movement and member of the Swedish government.

the feelings they both had. If Myrdal actually had the feelings he purports to have had, then nothing more is required. It is an entirely separate question whether our memories of our feelings long ago are reliable. (It is well known that often they are not, but in this example I neglect this aspect.)

The question regarding what was actually the case is often connected to the question of what is responsible for Jan Myrdal's feelings of abandonment? Was it his mother's actions, or his own excessive need for attention?

The supplementary question 'Who is to blame?' shows that values play a substantial role when we describe social situations. It is yet another sign that one, on a conceptual basis, can hardly describe a social situation without making a value judgment. We usually distinguish between propositions regarding facts and propositions regarding values (propositions that express moral, judicial or aesthetic norms, as in when something is good, bad, right, wrong, beautiful, etc.) in that we usually say that propositions regarding facts are true or false, whereas propositions regarding values do not, at least on some views, have this property. Thus the conclusion is that in describing social relations, we often cannot distinguish between fact and value judgment on the basis that the concepts we use are more or less impregnated with values. The example of Jan Myrdal's relationship with his mother is typical in that the concepts used to talk about human relations, concepts such as 'unloving' and 'egocentric', are both descriptive and evaluative. In short, there hardly exist any concepts that are such that we can describe a social phenomenon without expressing values.

A means to attain the highest level of objectivity possible in describing social relations is to introduce a reference to a certain person's perspective. When we say that 'Jan Myrdal's mother neglected him' it sounds like an objective and, in principle, observable fact. But a more precise expression of the fact alluded to would clearly state that the situation regards particular experiences: 'Jan Myrdal experienced that his mother neglected him'. This reformulation is perhaps not a big step forward as regards determining the fact of the matter, but it does make clear the context of the situation; namely, how a relationship is experienced. Thus if someone says 'this is true for me', and by this means 'this is how I experience the situation', then there is nothing here to criticize, save perhaps that the expression hides an allusion to private experience. However, it should be noted that such propositions are true or false in the usual sense of 'stating it as it is'. Either it is the case that Myrdal had such an experience and the proposition is true, or it is the case that he did not and he has lied, or incorrectly remembered. If the latter, then the proposition is false. That it might be impossible for us to determine which is the case is irrelevant.

2.9 Knowledge of Religious Beliefs

It is sometimes claimed that one can indeed have knowledge of something without believing that something. An example of this is religious belief. Consider the belief that Jesus is God's son. One is here mixing together two very different things. If I say, 'I know that Jesus is God's son', then I cannot claim that I do not believe it, since 'I know' implies 'I believe'. Though, one can simultaneously hold that 'According to the Christian religion, Jesus is God's son' and 'I do not believe Jesus is God's son'. In other words, I can have knowledge of what the Christian religion claims and the various beliefs that Christians hold, without holding those beliefs myself. I do not then know that Jesus is God's son, but rather I merely know what certain other people believe.

The above example is merely a special case of the trivial fact that we can have (true) knowledge of what someone believes, even if the content of those beliefs is not believed by us, or are simply false.

2.10 Summary

The classical definition of knowledge is that it is a true justified belief. In other words, knowledge is a state of mind that is about the world, that it is true and such that good reasons can be given for the content of the belief. One may say that the concept of knowledge connects the subjective, the intersubjective and the objective aspect of a state of mind.

We accept in all empirical sciences, in everyday life and in the courts that a piece of evidence could be very good reason for a belief without providing a complete guarantee for truth. Hence, since there is no strict connection between good reason and truth, it may happen that we have very good reason for a false belief. And further, Gettier discovered that we may have good reasons for a true belief, but we still do not want to say that that belief constitutes knowledge; we believe the proposition for the wrong reason, it seems. This has inspired many philosophers to suggest an alternative definition of knowledge, viz. that it is a true belief, which is arrived at by a reliable procedure. The point is that the person who is attributed a piece of knowledge may not know that he/she in fact used a reliable process. The reasons he/she may give for the belief is not relevant.

So what is knowledge? Philosophers disagree; but they do not disagree about the truth-condition; a false belief cannot be knowledge.

Discussion Exercises

1. Detective Carl Blomkvist is investigating a murder. At the scene of the crime he finds a wallet containing a driver's license that belongs to a Pelle Persson. The detective seeks him out and discovers that Pelle Persson left town in a hurry. It just so happens that Pelle Persson has long been in conflict with the victim and had much to gain from the victim's death. Pelle Persson is arrested; under his

fingernails the police find traces of the victim's blood. The police also find a large number of fingerprints belonging to Pelle Persson at the scene of the crime. There are no other suspects; and so Carl Blomkvist becomes convinced that Pelle Persson is the murderer.

In fact, Pelle Persson is not the murderer. Another person is, and he has succeeded in simultaneously hiding his involvement and framing Pelle Persson.

Does Carl Blomkvist know that Pelle Persson is the murderer?

2. In the 1500s, the geocentric worldview was still the established view. Among other things, this worldview enabled seafarers to determine their latitude when sailing on the open ocean by reference to the positions of the sun and other celestial bodies. Even though their worldview was not correct, their latitudinal calculations were.

   When such a seafarer correctly believes that his latitude is x degrees on the basis of the instruments and techniques of the day, does he know it?

3. Lady G is on a walk in the inner city of London and is thinking about classic rock'n roll. She sees a man on a street corner that she has seen on TV many times, as well as in the newspapers. However, he looks a bit older than he did in the pictures she saw. She stares intently at the person and says to herself, 'My God, that is Elvis Presley! I saw him!'

   In actuality, what Lady G tells herself is true. Elvis Presley is in fact living a quiet life in London, and it was just this person that she has met. However, at the time of this occurrence there is an Elvis Presley look-alike competition going on in London, so the city is crawling with Elvis impersonators. This event is well known and all over the tabloids, but Lady G does not read those papers and has no idea that the competition is taking place.

   Does Lady G know that she saw Elvis Presley?

4. Doctor H prescribes medicine M to a patient P. This medicine has been carefully tested and has been used for 20 years with a 100 % success rate, and no side effects. Doctor H knows all of this when she writes out the prescription; she claims that it will cure the patient and it does. However, no one knows how medicine M works. It is a complete mystery.

   Does Doctor H know that the patient will be cured?

5. Discuss the following propositions! Do you agree?

   (a) A proposition can be true even though everyone thinks that it is false.
   (b) A proposition can be true even though no one can give reasons for believing it.
   (c) A proposition can be false even though one has good reasons for believing it is true.
   (d) If one has good reasons for believing a proposition, then one knows it.
   (e) If everyone considers a proposition to be true, then it is true.
   (f) Many propositions that were true 500 years ago are now false.
   (g) If one knows that a proposition is true, then it is true.
   (h) If one knows that a proposition is true, then one believes it is true.
   (i) If one believes that a proposition is true, then one knows that it is true.

(j) 500 years ago, everyone knew that the earth was the centre of the universe, but now we know that the earth is not the centre of the universe.
(k) There is nothing that we know for certain.

Further Reading

Floridi, L. (2011). *The philosophy of information*. Oxford: Oxford University Press.
Hospers, J. (1986). *Introduction to philosophical analysis*. London: Routledge.
Kornblith, H. (2002). *Knowledge and its place in nature*. Oxford: Clarendon.
Longino, H. E. (2002). *The fate of knowledge*. Princeton: Princeton University Press.
Plato. (ca. 400 BC). *Theatetus*. http://www.gutenberg.org/ebooks/1726
Williams, M. (2001). *Problems of knowledge. A critical introduction to epistemology*. Oxford: Oxford University Press.
Wiseman, R. (2011). *Paranormality*. London, Basingstoke and Oxford: MacMillan.

Chapter 3
Hypotheses and Hypothesis Testing

We and other animals notice what goes on around us. This helps us by suggesting what we might expect and event how to prevent it, and thus fosters survival. However, the expedient works only imperfectly. There are surprises, and they are unsettling. How can we tell when we are right? We are faced with the problem of error.
W. V. O. Quine

3.1 Introduction

What does it mean that an activity, or argument, is scientific? Is there any single criterion that the natural sciences, social sciences and humanities must fulfil in order to be called science? Many have answered 'no', on the basis that the natural sciences and the human sciences are fundamentally two different enterprises. One central argument for this view, offered by Dilthey[1] and Weber,[2] is that the natural sciences and the cultural sciences (which was their term for human and most social sciences) have different purposes: the purpose of the natural sciences is the explanation of natural phenomena, whereas the purpose of the cultural sciences is the understanding of human phenomena. Weber may be interpreted as holding that explanation is causal explanation and understanding is discerning the meaning of actions of individuals and collectives.

There are further differences between natural and cultural science, which we shall discuss in detail in Chap. 5; but despite the differences there are also similarities. In my opinion, there is a common denominator, a defining trait, of all that is properly called science, and it is our task here to identify it.

Is there a point to finding a good definition of science? Yes, and for at least two important reasons. Firstly, there is a need to distinguish between science and pseudo-science: the latter being activities that claim to be scientific but upon closer analysis prove to be based on wishful thinking, fraud, business interests or pure

[1] Willhelm Dilthey (1833–1911), German philosopher and historian who helped develop German idealism. He saw cultural phenomena as objectifications of people's mental lives.

[2] Max Weber (1864–1920), German sociologist, historian and philosopher. According to Weber, the purpose of sociology is to understand the meaning and/or values expressed in human action.

L.-G. Johansson, *Philosophy of Science for Scientists*, Springer Undergraduate Texts in Philosophy, DOI 10.1007/978-3-319-26551-3_3

superstition. In particular, it is essential for the medical field to be able to clearly distinguish between medical science and quackery. In public debate, success stories are often reported regarding things like homeopathy, herbal remedies, healing, etc. Even if one may sometimes believe in these success stories, the central question is whether that success was the result of the remedying method or some extraneous circumstances. Thus it can be of great value to be able to distinguish scientific methods from non-scientific, as only with scientific methods is there a reasonable chance that the conclusions one draws are true and can be generalized.

The other reason for discussing criteria for science is to make possible an understanding of the immense transformations that we call the scientific and industrial revolutions. These revolutions began a line of research that has since then accelerated to an amazing pace. If we can find a satisfying characterization of what science is, then we can better understand how the propagation of science works, as well as further our understanding of a central component of our own society's transformation.

3.2 Unity of Science?

What is the common characteristic of all the activities that we call science? The naive answer would be that science discovers new things and increases our knowledge. An immediate rebuttal to this characterization is the fact that, in the future, presumably some of our present scientific beliefs will sooner or later prove to be mistaken. We cannot know *now* what is amiss with our present beliefs (for then we would not believe they were knowledge); but it is quite reasonable to believe that at least some part of what we now believe will later turn out to be false. If we claim that the criterion for scientific research is that it increases our knowledge, we would be forced to say that some of what we now call research, though we do not know which part, is not actually research (since it does not guarantee knowledge). However, to condemn some part of scientific research after the fact is not particularly productive. What we need is a criterion for science and research that we can use without it being necessary to know exactly what is true or false. Therefore, the most common view among philosophers is that criteria for science should be formulated in terms of *methods* and not in terms of *results*. This means that a certain enterprise can count as science even though it later turns out that the theories this enterprise put forward are not quite correct. It also means that an enterprise that results in true propositions is not necessarily scientific; they could be arrived at merely by coincidence.

We come thus to the following central question: is there some common method used by all the sciences that can stand as a criterion, or even *the* criterion, for science? In response to this question, one can identify three possible answers.

Some rather cynical people would answer no: there is no such common property that distinguishes science from other activities. According to this cynical view, what is called 'science' is arbitrary and the label is primarily used for enhancing the prestige of the enterprise so labelled.

According to another rather common view, inspired by Weber, scientific research is that which is typically practiced in universities, and there need be no significant similarity between all of these research enterprises beyond this. There is one crucial difference between the *natural* and *cultural* sciences, which is that the natural and cultural sciences use essentially different methods. The central activity of the cultural sciences is interpretation: interpretation of texts, events, actions, assertions, institutions, artefacts, etc. Such interpretation always contains a subjective element, as any interpretation must be based on the interpreter's prior understanding and cultural background. This in turn implies that interpretation is, in principle, not totally objective: i.e., not independent of individual perspective and culture. In contrast, in the natural sciences it is possible to be strictly objective in describing the phenomena one is studying in that two researchers can agree on an observation, irrespective of the cultural or theoretical background they might have.

Natural and cultural science are, according to this view, essentially different, and the fact that we call both types of enterprise 'science' does not imply that they share a method in common. On this view, those enterprises that do not fall into one of these groups, each of which is characterized by its methods, are deemed pseudo-sciences and hence are not sciences at all.

According to a third much disputed view, there is a least common denominator for *everything* that deserves the labels 'science' and/or 'research'. This is my view. The common denominator is the *hypothetical-deductive method*. Every enterprise that has the right to call itself science applies some variation of this method, though the details may differ from case to case. The hypothetical-deductive method, according to this view, should be used as the criterion for science; if an activity, taken in its entirety, does not abide by the requirements formulated in the hypothetical-deductive method, then it should be classified as a pseudo-science. I favour this third view.

Against those who claim that the cultural and natural sciences are fundamentally different, one may reply that the core activity of the major part of the cultural sciences–interpretation of texts, actions, and historical events–can be viewed as a sort of hypothesis testing where the interpretations play the role of hypotheses. These 'hypotheses' are tested against the types of evidence forthcoming in the texts, events, artefacts, etc. I shall explain this idea further in Sect. 3.5.

3.3 Hypothetical-Deductive Method

In order to more precisely describe what the hypothetical-deductive method is, I shall discuss a couple of examples from the history of science.

Example 1. The rejection of abiogenesis

For a long time people thought that worms, larva, etc. came into being spontaneously, rather than through the reproductive activity of parents. The idea was that these types of organisms were generated by decaying plant and animal matter. This idea is not entirely unreasonable given everyday observations of the natural world. Indeed, if one digs up a shovel's worth of topsoil, which is composed of plants in various stages of decomposition, one finds that it contains a large variety of small animals such as centipedes, spiders, larva, etc. (the existence of millions of organisms invisible to the human eye was, of course, not known in the mid 1600s or earlier). This view – theory is perhaps too pretentious a word – is called abiogenesis and was widely accepted up until the mid 1600s. However, a doctor from Florence named Francesco Redi (1621–1697) came to doubt abiogenesis. He tells (1668) how he

> . . . .began to believe that all worms found in meat were derived directly from the droppings of flies, and not from the putrefaction of the meat, and I was still more confirmed in this belief by having observed that, before the meat grew wormy, flies had hovered over it, of the same kind as those that later bred in it. Belief would be vain without the confirmation of experiment, hence in the middle of July I put a snake, some fish, some eels of the Arno, and a slice of milk-fed veal in four large wide-mouthed flasks; having well closed and sealed them, I then filled the same number of flasks in the same way, only leaving these open. It was not long before the meat and the fish, in these second vessels, became wormy and flies were seen entering and leaving at will; but in the closed flasks I did not see a worm, though many days had passed since the dead flesh had been put in them. . . . .
>
> To remove all doubt, as the trial had been made with closed vessels into which the air could not penetrate or circulate, I wished to attempt a new experiment by putting meat and fish in a large vase closed only with a fine Naples veil, that allowed the air to enter. For further protection against flies, I placed the vessel in a frame covered with the same net. I never saw any worms in the meat, though many were to be seen moving about on the net-covered frame.' (p. 33 ff.)

After these experiments, Redi (and eventually many others) became convinced that worms did not spontaneously arise (However, people continued to believe in abiogenesis of microscopic organisms until Pasteur successfully falsified this view).

The structure of Redi's argument is as follows: Redi begins by considering abiogenesis as a hypothesis (H), which he proposes to falsify with an experiment. We can formulate it as

(H) 'Worms arise spontaneously, in the absence of living animals, when meat and fish decompose.'

This is apparently one part of the abiogenesis view. From this hypothesis he draws the following conclusion:

(E) 'If I place meat and fish to rot in a covered jar, after a time I will be able to observe worms in the meat and fish.'

We call this sentence an Empirical Consequence (E) since, (1) it is a logical consequence of the hypothesis, and (2) its truth value can be decided by empirical experiments: i.e., we can prepare a situation such that we can show, through observation, whether the sentence is truth or false. Since the experiment shows the E is false, we must infer that the hypothesis is also false; because E is a direct consequence of H. The reasoning can be formulated thus:

If H, then E
E false

H false

This is the age-old way of a displaying a logical argument; premises go above the line and the conclusion is placed below.

Once we have formulated an argument in this fashion, it is easy to see whether the argument displays *logically valid reasoning*: that regardless of what the letters H and E stand for, the conclusion follows from the premises. Put another way, an argument is valid only on the basis of its *logical form*, which in this case is called *modus tollens* (see Appendix).

Have we now definitively proven that H is false? No, we have not. That the form of an argument is valid means only that *if* we accept the premises, then we must accept the conclusion. Thus we can still doubt the conclusion of a valid argument by doubting the premises. In particular, if we read the latter part of Redi's text, we see that there is a possible objection to H. When Redi acknowledges that further experiments are required to check that excluding the air in his initial experiments did not unduly inhibit larval growth, he is indirectly saying that one can make the following objection to his initial experiment: it neglected a necessary condition for the production of larvae from rotting meat, namely the free circulation of air. This condition is not met in the first experiment. In this experiment, Redi tacitly assumed that air circulation was inessential for the production of larvae in rotten meat. Such a tacit assumption is commonly called an *Auxiliary assumption* (A). Thus it so happens that there is a hidden premise in the argument:

(A) The production of worms in rotten meat and fish does not depend on air circulation.

This auxiliary assumption is an essential step in the argument; and thus we modify our logical formulation to include it:

If H and A, then E
E false

Either H or A is false

Consequently, H can no longer be disproved by the falsity of E: it is possible that H is true and A is false. It was in order to exclude this possibility that Redi conducted the second experiment, in which the auxiliary assumption is not used.

This experiment is a beautiful example of an application of the hypothetico-deductive method. The term itself was not invented until modern times, but this way of thinking had already begun to be applied during the scientific revolution.

We have here seen that it is possible to falsify a hypothesis given that one has identified, and controlled, all auxiliary assumptions. But can one definitively confirm a hypothesis? This question leads directly to the next example.

Example 2. The link between diet and urine pH.

Claude Bernard (1813–1878)–a physiologist from Paris–conducted experimental studies of various physiological properties of mammals. He discovered that carnivores have sour and clear urine, while herbivores have alkaline and turbid urine. This apparent rule held without exception until he one day observed a deviation among some rabbits that had recently arrived from a breeder. These rabbits had sour and clear urine, even though rabbits are strictly herbivores and so should have alkaline and turbid urine by the apparent rule. Thus there must have been something special about these particular rabbits. This prompted Bernard to make the assumption that these rabbits had been starved during transport to his laboratory. Starvation causes the body to break down its own muscle tissue, which is the same as eating meat on the metabolic level and hence would account for the clear and sour urine.

In order to test his hypothesis, Bernard performed a number of experiments. The first was to feed the rabbits their normal diet and observe their urine. The result was that the rabbits' urine became alkaline and turbid again. Then he proceeded to observe the urine of starved horses and found that they also had clear and sour urine. Bernard later dissected the starved rabbits in order to see if they had indeed metabolized their own meat. After these various investigations, all of which confirmed his predictions, he was prepared to accept the hypothesis and claim that he had found the explanation to why the newly bred rabbits produced uncharacteristic urine.

Now let us analyse the structure of this argument. It is obvious that the hypothesis that Bernard decided to test is the following:

(H) Herbivores produce clear and sour urine when they fast.

This hypothesis was tested with respect to a general rule, or perhaps we should call it a theory; namely, that herbivores' urine is alkaline and turbid under normal circumstances, whereas carnivores' urine is clear and sour. This supporting theory is required in order to draw to any testable empirical implications from the hypothesis. As in the foregoing example, we can call this assumption, untested in this experiment, an auxiliary assumption:

(A) Herbivores' urine is normally alkaline and turbid, and carnivores' urine is normally clear and sour.

From these two premises, H and A, one can draw the following empirical consequences (and probably many others):

(E1) Rabbits with clear and sour urine produce alkaline and turbid urine when fed grass.

(E2) Horses who are starved produce clear and sour urine.

(E3) The inner organs of rabbits with clear and sour urine show signs that their bodies have metabolized their own muscle tissue.

All consequences could be observed to agree with his observations. Could we now say that Bernard has *proven* his hypothesis? The answer will depend upon what we mean by 'prove'. If by 'proof' we mean a definitive argument such that doubt of its premises and conclusions will never arise, then we cannot say that Bernard has proven his hypothesis. Another experiment with different herbivores could turn out differently and disprove the hypothesis. This is a general, purely logical, conclusion from the fact that the hypothesis–like every other scientific hypothesis–states something about all objects of a certain kind, while even the most thorough series of experiments can only cover a sample of that kind. In this case, Bernard's hypothesis is about all herbivores, though only a very few individuals of this kind were examined by him. Absolute certainty, even if a great number of experiments all agree with our hypothesis, cannot be had. What we eventually can say, at most, is that we have good reasons in believing that our hypothesis in correct. We can summarize the foregoing argument in the following way:

If H and A, then (E1, E2 and E3)
E1, E2 and E3 are all true

H is supported

I have drawn a dotted line (instead of a solid line as in the last case) under the premises to indicate that the inference is <u>not</u> logically valid. Thus, we cannot state that H is true; rather, only that H is supported to some degree: that it now has more credence than it initially had.

If we compare the two examples, we see an asymmetry; it is possible to absolutely falsify a hypothesis, but it is not possible to absolutely prove one. Indeed, we learned this earlier. There are many examples in science where a generally accepted theory had to be given up based on more careful experiments, even when the reasons for believing that theory for a long time were very good.

It is time to summarize what the hypothetical-deductive method amounts to:

- Put forth a hypothesis.
- Infer an empirically testable claim from the hypothesis and eventual auxiliary assumptions.
- Determine the veracity of the empirically testable claim via experiment and observation.
- Depending on whether the empirical implications are true or false, determine whether the hypothesis is <u>supported</u> or <u>falsified</u>.

There are complications with this model to which I shall soon return. Presently it is pertinent to present definitions of the key terms:

Theory$=_{\text{def.}}$ A set of statements whose relations are explicitly stated.

Hypothesis$=_{def.}$ A statement that (i) we are not entirely certain about, and (ii) which is used as a premise in inferring empirical consequences.

Empirical Consequence$=_{def.}$ A statement that (i) follows from the hypothesis and eventual auxiliary assumptions, and (ii) whose truth, under plausible circumstances, can be determined by observation.

Auxiliary assumption$=_{def.}$ A statement that (i) is necessary in order to infer an empirical implication from a hypothesis, and which (ii) is not tested in the given situation, but is rather assumed to be true.

Auxiliary assumptions can be of different kinds: results from other researchers, relevant everyday observations, assumptions about the working of measurement devices, or even whole theories. All can be called auxiliary assumptions, if they are used when inferring testable claims from a hypothesis. Sometimes the matter is so obvious that no one notices that there is an active hidden assumption until closer inspection of the argument. Thus, when analysing an argument it is sometimes useful to first identify the hypotheses and the empirical consequences. Once one puts these side by side, it is often the case that one notices a logical gap between the hypothesis and its supposed empirical consequences. One may then formulate the propositions needed to fill that gap, thereby making explicit the auxiliary assumptions on which the argument is built.

3.4 Hypothesis Testing in the Social Sciences

Many have claimed that the formulation, and subsequent testing, of hypotheses is unique to the natural sciences, and that the social and human sciences[3] use altogether different methods. Without going too deeply into this discussion, I would still like to give an example from the social sciences that fits quite well with the model for hypothesis testing previously described. It is contained in Emile Durkheim's book *Le Suicide (1897)*, where he discusses the social factors that govern incidence of suicide. One of his ideas was that suicide incidence is connected to the strength of social connections that exist between people. He called this strength of inter-personal connection within a society that society's *degree of integration* and he assumed that the higher the degree of integration, the lower the risk that one looses the desire to live and commits suicide. But how does one measure degree of integration in a society?

Durkheim thought that there was a difference in the degree of integration between catholic and protestant societies. Catholicism and Protestantism have

[3] The traditional English label for the humanistic disciplines is 'arts and humanities', but those who, like the present author, stress similarities among different sciences have began use the label 'human sciences' instead. In earlier times English used the term 'moral sciences' as contrast to 'natural science'. In German one usually distinguishes between 'Naturwissnschaften' and 'Kulturwissenschaften', i.e. using the word 'wissenschaft' as a general label for all systematic study at universities.

very different views as to the role of the church as a link between God and humans. There are many more church activities within Catholicism, and these have the effect of bringing people together, which culminates in a higher degree of integration than in Protestant communities (It should be kept in mind that it is the societal effects of Catholicism and Protestantism which are relevant, not any theological differences.). This assumption allows for the possibility of making comparisons. For example, in Switzerland many cantons are broadly similar in all respects except whether Catholicism or Protestantism predominates. If Durkheim's hypothesis is correct, then the suicide rate should be lower in the Catholic provinces than in the Protestant provinces. This was, in fact, shown to be the case via statistical data. We can reconstruct Durkheim's argument in the following way:

(H) The suicide rate is higher in societies with a low degree of integration than in societies with a high degree of integration.
(A) Catholic societies are more integrated than Protestant societies, all other circumstances being equal.
(E) Catholic societies have lower suicide rates than Protestant societies, all other circumstances being equal.

E is thus a prediction and we may compare with the actual records:
Observations:

Society type	Incidents of suicides (per million)
Catholic provinces in Switzerland	86.7
Mixed provinces	212
Protestant provinces in Switzerland	326

One has to admit that the empirical implications agree with these observations. Naturally, this does not prove that Durkheim's hypothesis is correct, since there could be many other factors involved in determining incidents of suicide. However, the observations strengthen his hypothesis.

3.5 Hypothesis Testing in History: The Wallenberg Affair

Testing of hypotheses is commonly viewed as completely distinct from methods in the human sciences, which all are different versions of interpretation of human artefacts. I will here exhibit a historical case, highly interesting in itself, which very clearly exhibit a structure that nicely fits into the schema of the hypothetico-deductive method.[4]

[4] I have not reconstructed the argument so as to exactly fit into the schema of HDM, it would make the section much less readable. But I hope the reader with only little effort is able to sort out the different points, in particular the auxiliary assumptions to the different hypotheses.

Raoul Wallenberg–a Swedish diplomat active in Budapest in 1944 whose job it was to help Hungarian Jews escape the Holocaust–was arrested by Russian troops when they arrived in Budapest. On January 16, 1945, it was reported that Wallenberg and his property had been taken into Russian 'protection'. On August 18, 1947, the Russians sent a note (after inquiries from the Swedish government) that stated that Wallenberg was not in the Soviet Union. Some years later, the Swedish government made another inquiry as to Wallenberg's whereabouts, since rumours that Wallenberg was alive and in the Soviet Union were circulating. These inquiries resulted in a memorandum that was sent to ambassador Sohlman on February 6, 1957, from Deputy Foreign Minister Gromyko:

> ... In this matter, Soviet authorities have made the appropriate page-by-page review of archived records from assistant departments in certain prisons. As a result of this review, an archived document from healthcare services at Ljubljanka prison has been found, in which there is evidence that Raoul Wallenberg was admitted there. The document is in the form of a handwritten report – addressed to former Soviet Union Minister for State Security Abakumov and written by the chief of healthcare services at this prison, A.L. Smoltsov – containing the following information:
>
> 'I report that the prisoner of your acquaintance, Walenberg, died suddenly last night in his cell, apparently as a result of an induced myocardial infarction.
>
> In following your instructions to personally oversee Walenberg, I request instruction as to who will be responsible for performing an autopsy to determine the cause of death.
>
> 17.7.1947
> Chief of the prison's sanitation department
> Colonel of Medical Services – Smoltsov'
>
> On this report there is the following handwritten signature from Smoltov: 'Have personally informed the minister. Order had been given to cremate without autopsy. 17.7. Smoltsov.'
>
> There has been no more success in finding any other information in the form of documents or testimony, given the death of aforementioned A.L. Smoltsov on May 7, 1953.
>
> On the basis of what has been found, the conclusion drawn here is that Wallenberg died in July, 1947.
>
> Raoul Wallenberg was apparently arrested together with other prisoners in an area of Soviet war activity. At the same time, one can be certain that Wallenberg's later detention and the false information about him being sent to the Soviet Foreign Ministry for a number of years by former leaders of security agencies was the result of Abakumov's criminal activities. In response to his serious crimes, which aimed to cause all sorts of damage to the Soviet Union, Akabumov was, as you know, sentenced to death by the USSR Supreme Court.
>
> The Soviet government expresses its sincerest condolences in light of what has occurred and its deepest sympathies to the Swedish government and Raoul Wallenberg's relatives.

At the time of this letter, Gromyko's statement was generally accepted by the Swedish government and the Swedish people. However, over time this opinion changed and doubt of the truth of his statement gained traction. The reason for this was that reports began to come in from various freed Russian prisoners that Raoul Wallenberg was alive much later than the initial report indicated. Among these prisoners were four that had been stationed at Wladimir prison who stated that Wallenberg was there in the mid 1950s. These four prisoners included an Austrian whose name was not given, a Swiss named Emil Brugger, and two Germans named Horst Theodor Müller and Gustaf Rehekampff. The Austrian is the only person who purported having personal contact with Wallenberg. Brugger said that he had been in the stockade with Wallenberg and the other two said that they heard from other prisoners that Wallenberg had been in Wladimir. In a letter sent to the Soviet government on July 17, 1959, the Swedish government wrote, 'Of course the Foreign Ministry must attach great importance to independent testimony of such a precise nature regarding Wallenberg's presence in certain prisons during certain years in the 1950s.'

What is the truth? Hans and Elsa Villius consider in their book *Fallet Raoul Wallenberg* ('The Case of Raul Wallenberg') three hypotheses:

H1. Smoltsov's report is correct and RW died July 17, 1947.
H2. Smoltsov's report is correct insofar as it describes death of a certain person, but it was not RW but another one, with an almost similar name.
H3. Smoltsov's report is a fabrication made by Soviet authorities.

The couple Villius conclude in their book that there is overwhelming evidence for H1, that Wallenberg actually died in Ljubljanka prison in 1947. They base this position on a critical analysis of the Russian letter, and transcripts of the hearings of the four persons who claimed to have certain information about Wallenberg at Wladimir prison, and a great deal of other information that I have omitted here for lack of space. Some of the arguments resulting from their analysis are the following:

- The only document regarding Wallenberg that was found in the Soviet archives was Smoltov's letter. According to Kosygin, who was prime minister of the Soviet Union at the beginning of the 1960s, there is no dossier on Wallenberg in the Soviet government's archives. This is explained by the fact that Abakumov tried to dispose of all traces of Wallenberg after his death. This fact is strengthened by numerous accounts of former fellow prisoners who were questioned about their knowledge of Wallenberg by high security officers, and were thereafter moved, isolated and forced to guarantee that they would never mention Wallenberg. Such evidence agrees with the claim that Smoltov's letter is the only document found, since it was never sent from the medical department of Ljublanka and Abakumov did not know that it existed. It is also in accordance with standard procedure that Smoltsov wrote down instructions given to him orally by his superiors.
- That Smoltsov received instructions from Abakumov to personally oversee Wallenberg agrees with Abakumov's later actions: as a diplomat, Wallenberg was a particularly sensitive case.
- It also agrees with Abakumov's order to cremate without autopsy.
- It also agrees with the fact that the Soviet Foreign Ministry, in 1947, claims that Wallenberg was not in the Soviet Union: Abakumov simply lied when the Soviet Foreign Ministry inquired.
- It also agrees with the fact that Wallenberg's dossier could not be found in the Russian government's archives. That it no longer existed in the beginning of the 1960s could only mean that a high-ranking official in the Russian government

disposed of it (and Abakumov was minister for state security during the years in question).

- The misspelling of Wallenberg's name is easy to understand, since in Russian there is a tendency to pronounce double consonants singly. The misspelling is actually an argument for the documents authenticity; for Smoltov it is natural to write 'Walenberg' when writing to someone from which he had just received verbal instructions. He wrote the letter as he had understood the name in a verbal context.
- The four people who came forth with information about Wallenberg's stay in Wladimir prison during the 1950s can be doubted. The anonymous man (his name is known to the Swedish Foreign Ministry, but has never been published) from Austria lacks credibility: among other things, he has given different versions of his connection with Wallenberg, he testified long after the time of the event in question and he was supposedly a cell spy according to another testimony.
- Brugger has reported having contact with Wallenberg in an article in Berner Tageblatt. However, this story differs substantially from that told in front of a Swedish representative less than 3 weeks after the Tageblatt article was published. The differences heavily diminishes his credibility.
- The two Germans relate their stories to the same source, a Georgian named Simon Goguberidze, who in turn said that he heard people say that Wallenberg was in Wladimir. There are circumstances making it probable that Goguberidze is the source of Brugger's claims as well, in which case Brugger's, Muller's and Rehkampffs' claims all originate from the same source. We thus essentially have <u>one</u> third-hand testimony, and nothing more.
- A series of prisoners from the Ljublanka and Lefortovo prisons in Moscow, who in various ways came to know of Wallenberg, were later moved to Wladimir. They were there isolated for a time, since they somehow knew of Wallenberg. Yet, none of these people, who were all questioned after being released, claimed that Wallenberg spent any time in Wladimir during the 1950s.

There is another series of arguments for the claim that Wallenberg died in 1947. However, in summary we can say that the hypotheses (i) that Wallenberg died in 1947 in Ljublanka, and (ii) that Abakumov tried to dispose of any trace of him, are greatly strengthened by these circumstances.

If we instead try the contrary hypothesis that Wallenberg did not die in 1947, we must assume that either Smoltsov's letter was a fabrication, or that this letter is about another person. If it was a fabrication, then we need the auxiliary assumption that the Russians have constructed the letter with the intent of deceiving the Swedish government. In such a case, is it really believable that KGB–the Soviet security agency–would have chosen to construct a single unofficial document, rather than producing nothing at all? If one had wanted to deceive the Swedes into believing that Wallenberg was dead, would it not have been more effective to fabricate an official autopsy report, written by some deceased medical officer, instead of a handwritten note relaying orders for cremation without autopsy? Is it credible that KGB would have misspelled Wallenberg's name in such a fabrication? It seems the only reasonable answers to both questions is no.

If Smoltsov's note is about some other person, then there must have been someone besides Wallenberg, another sensitively handled prisoner, for whom Abakumov would require special attention, and with practically the same last name as Wallenberg. Is this auxiliary assumption really believable? No, it is quite improbable.

Elsa and Hans Villius thereby draw the conclusion that there is no reason to doubt the authenticity of the Smoltsov-document, and thus the hypothesis that Wallenberg died in Ljubljanka prison on the night of July 17, 1947, is strongly supported.

It can be added that the latest (last?) investigation of the case, carried out by a joint Swedish-Russian commission, came up with the same conclusion in January 2001 as had the Villius couple. The only significant difference between the two reports was some apparent evidence recorded in the commission's report that Wallenberg may have been poisoned and did not die a natural death.

3.6 Statistical Testing of Hypotheses

3.6.1 Bayesianism

To say that a hypothesis has been supported by an experiment is not very precise. It is, without doubt, desirable to have a measure for the credence a given hypothesis gains through experiment. Is it, then, possible to calculate the probability that a certain hypothesis is true, given a certain experimental result? This is a controversial question, involving the interpretation of *probability*.

The problem can formulated in the following way: we want to know the conditional probability (see Sect. 7.3) for a hypothesis being true, given a certain outcome of an experiment or measurement. Using the experiment's outcome, we can calculate this probability using Bayes' formula:

$$P(H|O) = \frac{P(H)P(O|H)}{P(H)P(O|H) + P(H_0)P(O|H_0)}$$

where P(H) is the initial probability for the hypothesis H, and P(H|O) is the probability for H given a statistical outcome O. Thus P(O|H) is the probability for the outcome O given that the hypothesis H is true, and $P(O|H_0)$ is the probability of an outcome O given that the null-hypothesis (the negation of the test-hypothesis) H_0 is true. I have here assumed that there are only two alternatives as regards hypotheses: the test-hypothesis and null-hypothesis. Investigations of the effectiveness of medicines against diseases provides good examples of this method. In such a case, we can formulate the hypothesis as

(H) The medicine has an effect on the progression of the disease.

The null-hypothesis is then,

(H_0) The medicine has no effect on the progression of the disease.

In order to calculate the probability of the hypothesis H, given an outcome O, we must determine the values of P(H), P(O|H) and P(O|H_0). The two conditional probabilities pose no fundamental problems, but how does one figure out the original probability P(H)?

If one interprets the concept of probability as a measure of the degree of subjective certainty–otherwise known as credence–one can imagine the following argument: We have a hypothesis H, which is true or false. If we do not have any specific information that argues for or against the hypothesis prior to our experiments, then we may assume that the initial probability that H is true is 50 %, and similarly that the probability of H being false (and thus the null-hypothesis being true) is also 50 %. We may assume that these probabilities represent our degree of belief in H and its contrary, and in the absence of any evidence one way or the other it is plausible that we have as much belief in one as in the other. Then we perform the experiments. With the help of Bayes' formula, we can now calculate the new probabilities; and thus generate a new probability distribution where the hypothesis is either strengthened or weakened due to its probability either going up or down, respectively. Furthermore, we now have a measure of the degree of strengthening/ weakening of propositions.

Contrarily, if one does not interpret probability as a measure of the degree of subjective certainty but rather as an objective property, then it is difficult to motivate the claim that the original probabilities for a hypothesis being true or false is 50/50. The actual unknown probability could be anything, and Bayes formula is useless if the original probability is not known.

A fundamental question is whether it is scientifically justifiable to assume that probability is a subjective phenomenon. Proponents of this methodology often argue that no matter what the initial probabilities are, after a number of experiments, the probabilities of different individuals converge. This means that despite the subjective nature of initial probabilities, the long run results are still, arguably, scientifically objective.

However, this view is not universally accepted. The majority claims that we are not justified in any assumption regarding initial probabilities of a given hypothesis. What remains then is to restrict oneself to calculating the conditional probability of obtaining a certain statistical outcome, given the hypothesis.

3.6.2 Statistical Inference -Neyman-Pearson's Method

The more common procedure in statistical testing is to assume the null-hypothesis and calculate the conditional probabilities of the outcomes obtained. The following (fictional) medical example may illustrate this approach. Suppose that a new medicine in the fight against AIDS has been created. Our hypothesis and null-hypothesis are the same as above:

(H) The medicine has an effect on the progression of the disease.
(H_0) The medicine has no effect on the progression of the disease.

(For the sake of simplicity, I will ignore the possibility of negative effects.) The null-hypothesis implies that the difference between the groups is exactly zero, which is not particularly reasonable. In a more realistic experiment, one would formulate the null-hypothesis in terms of the difference being less than some arbitrarily chosen value. However, I shall ignore this complication in this case.

In order to determine the possible effects of the treatment, nine patients were given the new medicine while nine other patients were used as a control group. The patients in both groups are chosen at random. After treatment the results were:

	Treated	Untreated
Alive	6	1
Dead	3	8
Sum	9	9

A statistical test shows that if the null-hypothesis were true, then the probability of the above outcome would be less than 1 %. Is this sufficient for abandoning the null-hypothesis, drawing the conclusion that the treatment is effective? Most would probably answer 'Yes', but where does one draw the line? How low a probability for the experimental outcome, conditional on the null-hypothesis, is required before one considers that null-hypothesis refuted?

The choice is determined by balancing two risks against each other: the risk of accepting a false test-hypothesis (by rejecting a true null-hypothesis) and the risk of rejecting a true test hypothesis (by accepting a false null-hypothesis). In practice, one often takes the threshold to be 5 %, 1 % or 0.1 %. If the probability of some experimental outcome, given the null-hypothesis, is at most 5 %, then one says that the result is one-star significant. If it is at most 1 %, then the result is called two-star significant; and similarly, 0.1 % is three-star significant. The methodological rule is that one should determine the level of significance before the experiment is done and then accept or reject the null-hypothesis as the case may be.

Plainly, this rule is not certain to give correct conclusions. The probability for the outcome, conditional on the null-hypothesis, can be higher than 5 % in spite of the null-hypothesis being false, and it is possible that the significance value is less than 5 % (or any other chosen limit) though the null-hypothesis is true. We talk about two types of errors:

1. We reject H_0 while in fact it is true: *error of the first kind.*
2. We accept H_0 while in fact it is false: *error of the second kind.*

Hence, making an error of the first kind means accepting a false test hypothesis, an error of the second kind means rejecting a true test hypothesis.

The conclusion of this discussion is that if we are not prepared to say that a hypothesis has some initial probability, then using statistical methods does not allow one to claim any more than in earlier examples: experimental results merely strengthen or weaken test-hypotheses in a rather imprecise and hand-waving fashion. However, if one allows that test-hypotheses do have known initial probabilities, then one can make precise the degree to which a test hypothesis is strengthened by some experimental result. For instance, in the immediately preceding example, if we, being subjectivists, assume that the probability distribution between hypothesis and null-hypothesis is 50/50 before the experiment, then we can calculate, using Bayes' formula, the hypothesis' probability given the above results to be 99 %. How sensitive is this probability for variations in assumptions about prior probabilities? Not much; if, for example, one is rather pessimistic and say that the prior probability for the hypothesis is only 10 %, its probability after the test is instead 87 %.

Comparing these two methods from a purely epistemological perspective, I see no substantial difference. Both contain an element of subjective decision. In both there is a risk of making a mistake, which as a matter of principle cannot be completely eliminated.

3.7 Unacceptable Auxiliary Assumptions: Ad Hoc-Hypotheses

As we saw in the Abiogenesis example, one can save a hypothesis from falsification if one can find an auxiliary assumption to which one can direct modus tollens. Recall that if two or more hypotheses are required in order to infer an empirical consequence, and if that empirical consequence is false, then one cannot logically determine which of the hypotheses are false. Thus, if one is loathed to reject a certain hypothesis, then it can be saved from falsification by adding new auxiliary assumptions. However, such an attitude is hardly conducive to good science; rather, one should be prepared to reject or revise even the most well established theories if there is enough evidence against them.

An auxiliary assumption that is proposed only to save another hypothesis from falsification is called an *ad hoc-hypothesis* (*ad hoc* = for this). It is obvious that one cannot accept ad hoc-hypotheses in scientific argumentation. Then how does one distinguish between acceptable auxiliary assumptions and unacceptable ad hoc-hypotheses? Karl Popper, among others, have argued that ad hoc-hypotheses differ from acceptable auxiliary assumptions in that it is possible to independently test the latter by conducting experiments with no connection to the original ones, but not possible to do the same for the former. A couple of examples from ancient astronomy should illustrate this idea.

Example 1 According to the ancient world-view, the Earth was the centre of the universe and all celestial bodies revolved around it in perfect circles at constant speeds. In the heavens, all motion was perfect. Observation of the sun, moon and the stars seemed to agree with this theory.

However, the motion of planets, such as Mars, does not fit the theory. Viewed against the fixed stars Mars sometimes appears to move 'backwards'.

Every day Mars appears to move one revolution around the Earth, so that against the fixed stars it moves slowly from west to east. However, sometimes Mars 'backs up' in what is known as retrograde motion, as can be seen in Fig. 3.1. Does this mean that Mars sometimes slows down? No, such imperfect motion was not reasonable for the ancient astronomers.

Ptolemy (Greek astronomer, active in Alexandria circa 90–168 A.D.) construed an improved system that agreed with the observed retrograde motions. He claimed that each planet moved in a small circle, an epicycle, around a mathematical point located on the planet's main circular path around the Earth. He also placed the Earth a little bit away from the centre of the main circular path. When you combine these two circular motions–the epicycle and the main circular path–with suitably chosen radii, you get the apparent motion of the planet. Seen from Earth, Mars is in retrograde motion when it moves in a direction opposite to that of its main path. Since one has no restrictions on the choice of the epicycle, one can construct a system that agrees perfectly with the observed retrograde motion. However, given that there is no argument for why Mars should rotate around a mathematical point in space and since there is no independent way to test the hypothesis, the assumption of epicycles is an ad hoc-hypothesis (Fig. 3.2).

Another problem in ancient astronomy was to explain the existence of two kinds of eclipses, total and annular. It is noteworthy that the apparent sizes of the sun and the moon, as seen from Earth, are very nearly the same; hence when the Earth, Moon and Sun are all aligned, the moon is just the right apparent size to hide the Sun from the Earth in what we know as an eclipse. If, as the ancients believed, the Sun and Moon's orbits of the Earth were perfectly circular, with the earth at the centre of that circle, then every eclipse would be the same. Yet, all eclipses are not exactly similar; some eclipses are total–where the sun is completely obscured from the earth–and some are annular–where one sees a thin ring of the sun past the edge of the moon. We now know that this difference in eclipses is due to our Earth orbiting the sun in an ellipse with the sun at one of the ellipse's foci; however, this explanation was not available to the ancients as such elliptical motions were incompatible with the their divinely perfect heaven; according to ancient views, an ellipsis is a non-perfect, misshaped circle.

One hypothesis that was proposed at the time to explain this conflict between theory and observation was that the moon sometimes shrinks and expands! But the change in size is so small that the only way to notice it is to observe its effect during solar eclipses, which implies that there is no way of independently testing this assumption. Thus, this assumption provides another example of an ad hoc-hypothesis.

However, if we consider this matter from a present day technological point of view, the situation is quite different, since nowadays it is relatively easy to

Fig. 3.1 *Retrograde motion. The motion of Mars* from late October 2011 (*top right*) through early July 2012 (*bottom left*) (Reproduced with kind permission of Tunc Tercel)

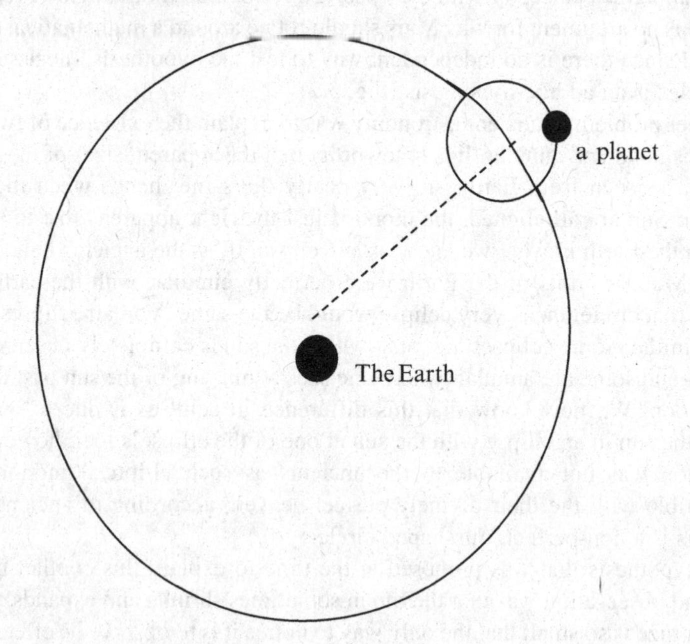

Fig. 3.2 The basic principle of Ptolemy's planetary system

accurately measure the diameter of the moon, for example using lasers. This shows that the question of whether the auxiliary assumption is testable or not is not so easy to determine as it might first appear. Criteria regarding independent testability do not seem to be absolute. Rather, they have to do with what is possible at a particular time. Even though all scientists have a strong intuition about what should count as an ad hoc-hypothesis, it is thus difficult to give a definite criterion.

3.8 Summary

The least common denominator for all sciences is that hypotheses are formulated and tested. This is meaningful only if one is prepared to change one mind after testing, to admit that even one favourite hypothesis was wrong. This state of mind is a crucial part of the scientific attitude.

Testing hypotheses against observations require auxiliary assumptions. They can also be tested, of course, although in a particular case of hypothesis testing they are taken for granted at the beginning.

The result of the test is either that the predictions and observation reports are compatible, or that they conflict. In the former case one may be justified to say that one's hypothesis is supported. In the latter case one must reconsider something; one must reject either the hypothesis, some auxiliary assumption, or the observation report. If one decides that the hypothesis is to be rejected, one has falsified it. It is thus clear that falsification of a hypothesis is no automatic inference from the test result; it is the result of a judgement, all aspects considered.

Exercises

Below are three short summaries of popular scientific articles. Give an analysis of the argumentation in each abstract using the concepts of hypothesis, auxiliary assumption and empirical consequence!

1. It has been generally assumed that the dinosaurs were cold-blooded animals. This assumption has been criticized by Stephen Jay Gould. He claims that they must have been warm-blooded because:
   (a) The temperature of cold-blooded animals' varies with outside temperature. Cold-blooded animals that live in areas with large changes in temperature between summer and winter get growth rings in the outer parts of their skeletons, similar to the growth rings of trees. Warm-blooded animals do not get such rings since they maintain, more or less, a constant body temperature. Dinosaurs that lived in areas of varying temperature did not have such growth rings.
   (b) Large cold-blooded animals do not live near the Polar Regions since they cannot get enough warmth during the short winter days and are too large to find secure shelter. Some large dinosaurs lived so far north that they must

have spent long periods of time without sunlight; and thus without an external heat source during the winter.

(c) Modern reconstructions of the anatomy of various dinosaurs show that many large dinosaurs are similar to present-day mammals (which are warm-blooded), both in regard to skeletal muscles and the proportions of various limbs.

2. The following text is an excerpt from the article 'The Confirmation of the Continental Drift' in *Scientific American*, April 1968. It discusses the possibility that Africa and South America were once joined, an idea that seems quite natural when looking at the contours of these two continents.

Of particular interest for us was the sharp borderline between the 2000 million year-old geological area in Ghana, the Ivory Coast and further west, and the 600 million year-old area of Benin, Nigeria and further east. This borderline runs southwest towards the ocean near Accra, Ghana. If Brazil (and all of South America) had been part of Africa 500 million years ago, the borderline between these two areas would have run through South America near the city of Sao Luis on the north coast of Brazil. Therefore, out first task is dating the areas near Sao Luis.

To our pleasure and surprise, the results fell into two groups: 2000 million years on the west side and 600 million years on the east side of the borderline where we had expected it to be. It appeared as if a 2000 million year-old piece of Africa had landed on the South American continent.

3. Over the years 1844–1848, Ignaz Semmelweis worked as an obstetrician in Vienna. The hospital where he worked had two maternity wards, each taking approximately 3000 admissions per year. It so happened that the proportion of deaths due to puerperal fever was markedly different between the two wards:

	Ward 1 (%)	Ward 2 (%)
1844	8.2	2.3
1845	6.6	2.0
1846	11.4	2.7

A great many explanations of this difference were proposed, such as incorrect birthing position, poor diet, 'atmospheric conditions' etc. All of these could be discarded quite reasonably. Then Semmelweis got the idea that some 'cadaveric material' from dead persons had been transported to Ward 1, which agreed with the fact that at Ward 1 medical students were instructed after having partaken in autopsies, while at Ward 2 midwives, who had no contact with cadavers, were instructed. As an experiment, Semmelweis had all who came to Ward 1 from an autopsy wash their hands in chlorinated lye. The results of this experiment on the number of deaths due to puerperal fever were the following:

	Ward 1	Ward 2
1848	1.27 %	1.37 %

(Unfortunately, Semmelweis' experiment was not accepted as a good argument for continuing to wash one's hands after autopsies. Semmelweis was dismissed and the use of disinfectants ceased. It took more than 20 years until the practice of disinfection became commonplace, thanks to Pasteur's discovery of bacteria.)

Further Reading

General

Hempel, C. (1966). *Philosophy of natural science*. Englewood Cliffs: Prentice Hall.

Hermeneutics as Hypothesis Testing

Chesterman, A. (2008). The status of interpretive hypotheses. In G. Hansen, A. Chesterman, & H. Gerzymisch-Arbogast (Eds.), *Efforts and models in interpreting and translation research: A tribute to Daniel Gile* (pp. 49–61). Philadelphia: John Benjamins.

Statistical Testing of Hypotheses: There Are Numerous Textbooks on Neyman-Person's Method, e.g.

Moore & McCabe. (2005, 2009, 2012, 2014). *Introduction to the practice of statistics*. New York: W.H. Freeman and Co, chapters 6 and 7.

Bayesian Inference, See e.g.

Kruschke, J. (2014). *Doing Bayesian data analysis: A tutorial with R, JAGS, and Stan* (2nd ed.). Burlington: Academic.

Chapter 4
On Scientific Data

It is therefore correct to say that the senses do not err – not because they always judge rightly, but because they do not judge at all.
Immanuel Kant: The Critique of Pure Reason

4.1 Measurement and Scales

Galilei, Kepler, and some other scholars around 1600 began making systematic measurements of the motion of bodies on earth and in the sky. This was a crucial aspect of the scientific revolution. At start such measurements was not consciously aimed at testing hypotheses, but the first results could be generalised to simple regularities, which could be treated as hypotheses and tested in new experiments in which kinematic quantities are measured.

A measurement is a comparison with a standard of some kind, and the result of repeated measurements is an ordering of objects in terms of how they compare to the standard. Another was of saying the same thing is to say that a particular type of measurement determines a scale, where each scales impose an order among a sets of objects. One says that the pike *weighs* 4 kg, that the A4-paper is 210 mm *wide*, or that Ann got an *A in English*. A statement about a measurement is an indication that a certain object is assigned a *measure* and a *quantity*. One can have various units for measurements of the same object. For example, length can be measured in meters, yards or feet. For it to be meaningful to say that one and the same property can be measured in various units, there must exist a well-defined transformation that converts one unit into another. The characteristics of these transformations can be used to give precise definitions of the different types of scales, as we shall soon see. First, however, I will give an informal characterization of four scale types.

A **nominal scale** (Latin nomen = name) is simply a classification of observations into categories. For such scales, no numerical comparison between values is possible, since no numbers can be assigned. One cannot say that some object has more or less of some property than another.

Example Consider the blood types A, B, AB, and O. One cannot say of one of these types that it has more or less of some property than any other. Blood type A has the A-factor but not the B-factor, and blood type B has the B-factor but not the

L.-G. Johansson, *Philosophy of Science for Scientists*, Springer Undergraduate Texts in Philosophy, DOI 10.1007/978-3-319-26551-3_4

A-factor. The only thing we can say is that two people with different blood types have blood with different properties.

An **ordinal scale** is a ranking of objects with respect to some property where objects have more or less of the measured property. However, there is no universal standard as regards to the relative sizes of the increments in the scale: all that can be said is that one object has more or less of some property than another, not how much more or less of that property.

Example Many grading systems are ordinal scales, such as the European ECTS grading system A, B, C, D, E, FX and F. In this system one can say that a student who was given a C is more knowledgeable (according to the teacher) than a student who received a D, and that a student who received a B is more knowledgeable than both. Yet, we cannot say that the difference between B and C is the same as the difference between C and D.

An **interval scale** is a scale where the measurement data is ordered into a hierarchy with equal steps between increments. This means that one can make quantitative comparisons of measurements.

Example The most common scale for measuring temperature is Celsius. Using this scale, we can show that the differences between 30 °C and 20 °C, and 20 °C and 10 °C, are the same; since a difference of 1° is equal for any two consecutive points on this scale. However, we cannot make sense of ratios of measurements in this scale, as this would presume a non-trivially chosen zero-point representing the absence of the property being measured. 0 °C is not the absence of temperature, but rather a zero chosen for practical reasons, viz., the freezing point of water.

Finally, the **quotient scale** is an interval scale with a fixed zero-point representing complete absence of the measured property. This allows meaningful talk of quotients between measurements of different sizes.

Example Measurements of length. An object's length is measured using a variety of instruments: meter sticks, callipers, lasers, satellites, etc. We have an internationally accepted scale of measurement (1 m = the distance light travels in 1/299,792,458 s). If an object has a length of 0 m, then it has no length. Thus we can talk about a certain object being twice as long as another. If the zero were arbitrarily chosen, then the ratios between measurement values would change depending on where one chooses to place the zero-point. By comparison, we can hardly say that 20 °C is twice as warm as 10 °C, as becomes obvious when one expresses the same temperature in the Fahrenheit scale: 10 °C = 50 °F and 20 °C = 68 °F; hence this ratio of 2:1 in Celsius is a ratio of 6.8:5 in Fahrenheit.

Thus a scale is an assignment of numbers to objects (one can even assign numbers to categories in a nominal scale). Such an assignment is called a *measure*. With this terminology in hand, we can now give the following formal definitions of the various types of scale:

Def. 1: A measure f for a quantity q is a *quotient scale* if and only if for any other measure g of the same quantity there exists a number $k > 0$ such that $f = kg$

Def. 2: A measure f for a quantity q is an *interval scale* if and only if for any other measure g of the same quantity there exists a number $k > 0$ and a real number r such that $f = kg + r$

Def. 3: A measure f for a quantity q is an *ordinal scale* if and only if for any other measure g of the same quantity there exists a strictly increasing function j such that $g = j(f)$

Notice that the nominal scale is not given a formal definition above. There is perhaps reason to claim that the nominal scale is not actually a scale in the full sense of the word, but merely a classificatory scheme. However, the reason why it appeared first in the informal characterization above is that, in the transition from the nominal scale to the quotient scale, via the interval and ordinal scales, one sees an increase in the structuring of one's set of observations. I shall discuss the different types of classifications, especially classifications in the humanities and social sciences, in more detail in the next chapter.

In order to facilitate automatic recording and statistical processing one often assigns reference numerals to recorded observations, but it is a mistake to treat these labels as *numbers*, which are the normal reference of *numerals*.[1] This situation often arises with surveys. Many different types of surveys are comprised of various claims and the one taking the survey is required to mark one of the following alternative responses to each claim: *Strongly Agree, Somewhat Agree, No Opinion, Somewhat Disagree, Strongly Disagree*. This is a typical ordinal scale. It is commonplace that one assigns to these alternatives the following numerals: 5, 4, 3, 2 and 1. But the use of numerals to designate outcomes does not imply that one is using an interval or quotient scale; indeed, in this case the scale is plainly ordinal. In such situations, where numerals serve merely to rank or label, calculating means and standard deviations over them is meaningless. For it to be meaningful to calculate means and standard deviations over the results of opinion surveys the difference between, e.g., 'strongly agree' and 'somewhat agree' would have to be the same as the difference between 'no opinion' and 'somewhat disagree', but as standardly conceived this is just not the case.

4.2 Statistical Relations

It is quite often the case that we want to know if there is some connection between two variables. In an ideal experiment, one variable is varied in order to study the effects on another while all other variables are kept constant. But in most cases this is not possible. In many disciplines, such as medicine or climate research, we just do not know all of the relevant variables involved in a certain situation. Furthermore, even if we knew exactly which variables should be held constant, it is often

[1] A numeral is a symbol, a linguistic sign that usually denotes an abstract object, a number. However, in some cases the numeral denotes a category of responses.

practically impossible to do so; how is one to construct an experiment to study the effect of a volcanic eruption on the world's average temperature? There may also be ethical restrictions, such as that one shouldn't manipulate people, animals or the environment without due regard to their wellbeing.

The conclusion is that actively controlled experiments are often infeasible, and that one must often be satisfied with passive observations instead. In order to determine whether two variables are in some way connected, one has to study many cases. The idea is that if one studies sufficiently many, randomly selected cases, then one has reason enough to believe that all other factors effecting the variables one wants to study are haphazardly distributed among the studied cases. Using statistical methods, one can then estimate the probability that an eventual connection between two studied variables is the result of the influence of uncontrolled variables; the larger the number of cases, the smaller the probability that the eventual connection does not obtain.

The first step is to conduct a sample test of a population and measure the two properties. One can then study the graph one gets if one places one of the variable properties on the x-axis and the other on the y-axis. Consider the following example:

A laboratory studied the relation between how long it takes for laboratory assistants to perform a certain procedure and the number of mistakes that were made. The results are presented in the following table.

Lab assistants	Time in minute	Number of mistakes
A	9	1
B	8	4
C	6	6
D	5	2
E	7	2
F	6	5
G	5	3
H	6	2
I	4	6
J	7	3

It is difficult to infer anything from this table, but let us look at the so-called scatter plot (Fig. 4.1).

In this diagram one can see a certain tendency that the longer a procedure takes, the fewer mistakes are made. A measure for this tendency is the so-called *coefficient of correlation* (also called *product-moment coefficient of correlation*) which is given by the formula

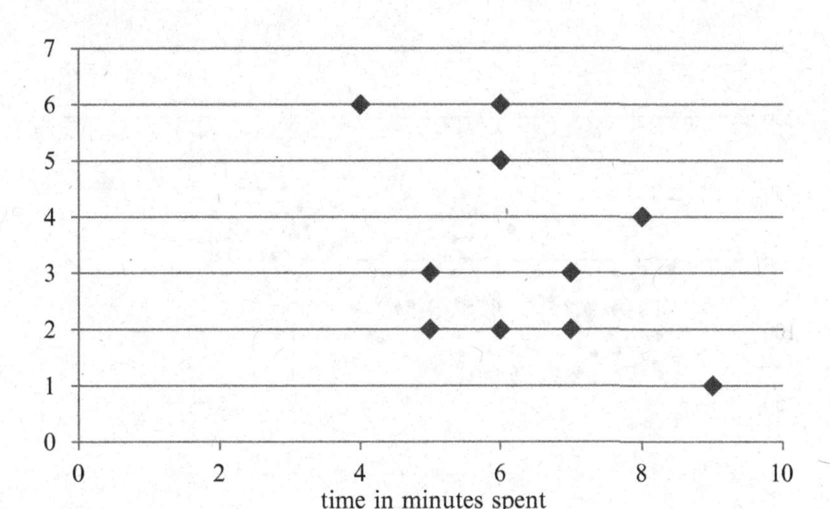

Fig. 4.1 Time spent versus number of mistakes in laboratory work

$$r_{xy} = \frac{n\Sigma xy - \Sigma x \Sigma y}{\sqrt{\left(n\Sigma x^2 - (\Sigma x)^2\right) \cdot \left(n\Sigma y^2 - (\Sigma y)^2\right)}}$$

where n is the number of observations, and x and y are the two variables. In this case the correlation is -0.47, which is a moderately negative connection. Of course, this was just a sample test, and without more information no definite conclusion can be drawn; it may very well happen that a larger sample test will show no correlation. But let us assume, for now, that this negative correlation is close to the actual: that it is the correlation that would result in an unlimited test. Given this assumption, our guess that the tendency for fewer mistakes to be made when time is increased is correct, since the correlation coefficient is negative.

Correlation coefficients range between +1 and -1. If the correlation is +1 we have a fully positive correlation, and if it is -1 we have a fully negative correlation. The minus sign shows that an increase in the value of one variable tends to decrease the value of the other variable. The following are two scatter plots with calculated correlations for different sets of data (Figs. 4.2 and 4.3).

The second scatter plot shows a 'thicker' figure than the previous, which implies a lower correlation. Roughly speaking, the thinner and longer the area covered by accumulated points is, the higher the correlation. In the next graph we see a correlation between data from two secondary school physics tests, of which one is called the 'practical test', and the other 'paper-and-pen' (Fig. 4.4).

Here we see that the plot is quite narrow, which implies a high correlation. The calculation $r_{xy} = 0.81$ confirms this. (I once conducted these two tests myself in order to see if there were any significant differences between the practical and

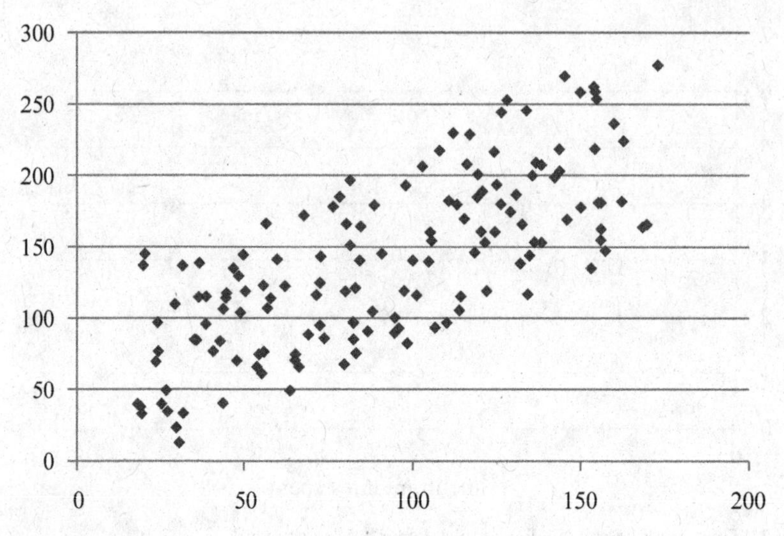

Fig. 4.2 $r_{xy} = 0.74$

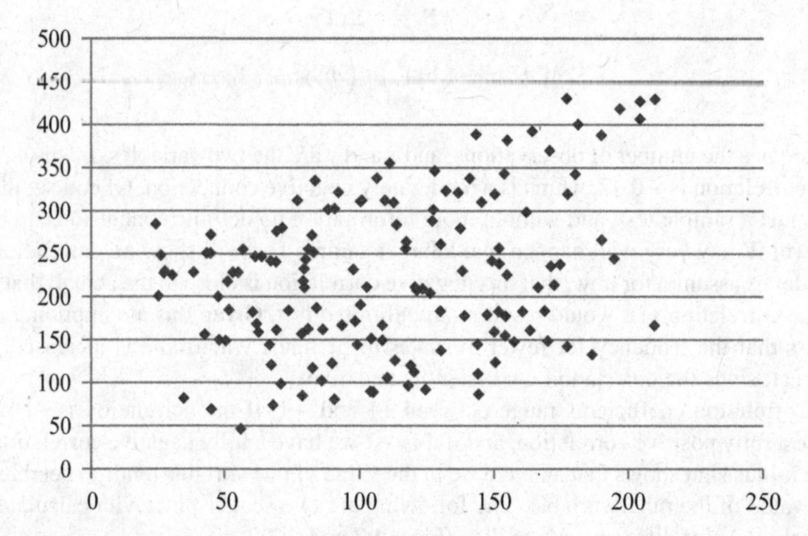

Fig. 4.3 $r_{xy} = 0.33$

theoretical capabilities of my physics students. The high degree of correlation shows that there was not.)

In the extreme case, when all points are located on a straight line, $r_{xy} = 1$ (or if negative, then $r_{xy} = -1$). The exception is when the line is vertical or horizontal, which means that the coefficient of correlation is 0.

However, the coefficient of correlation is not always useful as a measure of correlation. A moment's reflection tells us that this mathematical measure presupposes that both variables form interval or quotient scales, i.e. scales with

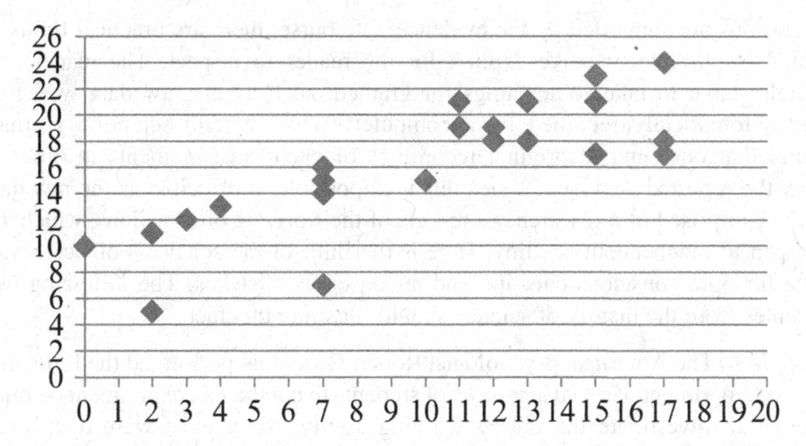

Fig. 4.4 The results of a paper-and-pen-test versus the results of an experimental test in a physics class

equidistant steps. If one of the scales is an ordinal scale, one must use other means for studying correlations.

In Chap. 7 we will discuss the conclusions one can draw from the existence of a correlation between two variables.

Exercises

For each of the following examples, state the type of scale the collected data will likely form.

1. An interview questionnaire containing the following question: How are you going to vote in the upcoming referendum?
2. A questionnaire containing the following question: How would you rate the healthcare system in Stockholm county: Very Good, Good, No Opinion, Bad, or Very Bad?
3. Classification of the colour of 20 particular species of forest mushrooms according to the international colour code.
4. pH measurements in the urine of a group of patients.
5. Measurements of temperature, in degrees Kelvin, of a number of stars in the milky way.

4.3 Data, Observation, Observational Statement

One of the central elements of science is the critical discussion of results; in any scientific paper all observations, conclusions, theories and hypotheses should be openly stated so that any competent reader is able to judge the extent to which the conclusions are supported by the evidence. Of course, there are practical limits to what it is possible to make explicit for the reader to inspect. The reader will naturally have to take some things for granted, such as the raw data, which is often automatically recorded by a computer. While certain scientific journals require that one send in original recordings of one's measurements in order to check the reported data, sometimes this is impossible in principle as the raw data maybe comprised of a researcher's reports of their private observations, which are not open to independent scrutiny. Here is the limit of the science's objectivity; a space for both conscious cheating and unconscious mistakes. The following two examples from the history of science should illustrate this fact.

Example 1 The American psychologist Robert Rosenthal performed the following famous experiment. He told a number of students to conduct an experiment on mice in order to investigate the mice's learning ability. Sixty mice were distributed among 12 students. Half of the mice were said to be of a particularly 'gifted' kind, and half were supposedly of a 'slow' kind. The students were to place the mice in a labyrinth in which one route led to a food bowl. The students would then study the mice's ability to learn by measuring the time it took for them to find the food. The result was as expected: the group of mice that were more 'intelligent' was quicker to learn than the not so gifted mice. In addition, the students noted that the quick thinking mice refused to move 11 % of the time, whereas the less gifted refused to move 20 % of the time.

The point of this experiment was not to investigate the intelligence of the mice, but to discover something about the students! In fact, there was no intrinsic difference between either of group of mice! The students had merely let their expectations direct their observations of how long it took the mice to find food. It may strike one as strange that there could be a subjective element in the measuring of time, but such is the case according to Rosenthal. His general conclusion, based on this experiment, was that there is a measure of unconscious judgment, not a simple recording, in each observational statement. Note, however, that this concept of *judgment* is different from the *interpretation* of intentional entities like human actions, artefacts and texts, to be discussed in the next chapter.

Example 2 For a long time surgical transplantation was plagued by the body's rejection of foreign objects. Then a young American scientist claimed to have succeeded in controlling the mechanisms of the immune system to prevent rejection of foreign objects and tissue. He presented a number of white mice with large black fur patches on their backs, which he claimed to have been transplanted onto the mice. Other researchers then attempted to duplicate this experiment, but all such attempts failed. Not until an assistant to the American scientist happened to get black paint on his fingers upon returning the mice to their cages did people begin question the validity of his results. In fact, it turned out that the scientist has simply painted black spots on the backs of the mice!

The blatant fraud committed by the scientist in the story of the mice with black spots triggers the question of how common such deceit really is? No one knows for sure; however, in their book *Betrayers of the Truth*, Broad and Wade discuss a large number of fraudulent cases, and come to the pessimistic conclusion that deceit in science is widespread. There are two explanations for this. Firstly, since academic careers are extremely competitive, and largely depend on the significance and quantity of the results one has published, there is a great incentive to cheat. As they say in USA, 'Publish or perish'. Secondly, little academic merit is accrued by repeating previously conducted experiments; so few experiments are ever repeated. Not until there is reason to suspect fraudulent or sloppy work does anyone take the trouble to repeat an experiment. Hence according to Broad and Wade there is both powerful motive, and ample opportunity, for fraud in the sciences.

One should perhaps not over-generalize Broad and Wade's conclusions. As they themselves point out, the possibility and risk for fraud is different in different scientific disciplines. In some disciplines, a great mass of empirical research is conducted in rather routine fashion. This is what Kuhn calls *normal science* (see Chap. 6). An individual investigation often aims at testing a fairly well established hypothesis, theory or method in a new situation, e.g. the measurement of physical or chemical properties of new substances. For the individual researcher, an investigation can often be motivated by the prospect of greater academic accreditation, while the relevance to the scientific community is often restricted to the results strengthening an already successful theory. Thus, scientific results can often glossed as 'we have confirmed that our methods, or theories, hold in yet another case'. In these kinds of situations there is a significant risk of fraud, as the chance of detection is minimal. However, it would be too hasty to draw the conclusion that fraud and deception are common practice in the scientific community, or that experimental results are entirely untrustworthy. My own view is quite the opposite; namely, that fraud and deception belong to the exception and not the rule. After all, just because scientists have motive and plenty of opportunity to commit fraud, this does not mean that most will do so.

Returning to the central theme of this section, we shall now differentiate *observations* from *observational statements*. Observations are mental processes; and as such, they are private and exempt from external verification. Observational statements – which include the statements made by scientists and researchers – are verbal or written and so externally verifiable in principle. Observational statements are what scientific theories are tested against, not the observations reported by such statements. To make an observational statement – to report what one has observed – is a more or less conscious human act, to judge, and not a passive reflection of incoming stimuli. The senses, in so far as we regard them as distinct from the mind, merely transmit signals to the mind. This is what Kant implied by the citation at the beginning of this chapter.

4.4 On the Theory-Dependence of Observational Statements

Intuitively, it seems an easy task to distinguish between observation and theory. A theory is something abstract, often comprised of a number of formulas or general sentences. (Recall the definition of a theory presented in Chap. 3; a theory is a set of propositions whose inherent connection is explicitly stated). An observation, on the other hand, is a concrete action; something one does. However, actions must be described using words if they are to be of any value in a scientific community; hence, observations are described by observational statements.

The question then arises: is it just as easy to distinguish between theoretical statements and observational statements? The answer is no, as can be seen from the previous examples regarding how unconscious background beliefs can affect what is observed and reported even a in a very simple tasks such as time measurements.

Many have drawn from such considerations the conclusion that all observation is theory-dependent. Their argument is as follows: Observation is an active process. Our sensory organs are exposed to physical stimuli, which spark a number of mental activities including processing and sorting of those stimuli in various ways. This processing is shaped by one's past experience, attentiveness, attitudes and expectations. The result is an observation expressed as an observational statement. Formulation of this statement requires a language containing a number of general terms by which one describes ones observations. This language and its categorizations are not theory-neutral; rather, they are mostly formulated in accordance with our preconceptions of the world. In short, our use of language is not theory-neutral in the strict sense of the word; neither the act of observation, nor the observational statements expressing those observations, are theory-neutral. N. R. Hanson argued this point in his (1965).

W. V. Quine disagrees[2] to a certain extent. He concedes that many of the observational statements in science are highly dependent on theory, but, he claims, there does exist some basic observational statements that everyone would agree to, independent of observers' culture, perspective or beliefs. A couple of obvious examples are 'the sky is now blue', 'there are seven apples in the basket', and 'the meter presently shows no deflection'. These statements describe here-and-now situations; and thus they are called *occasion sentences*. All competent speakers of the language in which these statements are expressed are able to immediately determine whether they are true or false simply by observing what is the case here and now. Thus according to Quine there is a solid foundation for our theorizing, though only some of the sentences we usually call observational statements are, in fact, theory-independent *occasion sentences*.

This opposing argument is based upon an important premise; namely, that two observers have access to the same language. This is certainly not always the case.

---

[2] This argument can be found in several of his writings, e.g. in the first chapter of his *Pursuit of Truth*, Harvard University Press, 1990.

There would be no issue if one could claim that, in principle, there is no problem in translating sentences of one language into sentences of another. Unfortunately, as any translator knows, this is far from the truth. Quine has discussed this question in connection with an investigation of the concept of *linguistic meaning*, and his conclusion is that there is a degree of indeterminacy, far from negligible, in any translation from one language to another.[3] Yet, no translation is entirely arbitrary: for, if a translation is to be acceptable, it must translate occasion sentences in one language held true by persons talking that language into occasion sentences of the other language held true by persons talking the second language.

In my opinion, Quine is correct. This foundation of occasion sentences whose truth-value people agree upon is what is needed in order to claim that science has any measure of objectivity. It should be observed that if two or more persons agree to assent or to dissent to a sentence, they need not interpret it the same way. One can for example imagine two persons with different world-views and different languages seeing a horse in the meadow. One has adopted Plato's world-view, and says, in his language, something like 'Lo, the Horse-form is instantiated', the other thinks in terms of nature populated with individual objects belonging to different categories and says, in his language 'Lo, a horse.' (I have here given a translation of both expressions to English, of course.) If the speakers agree on what they are looking at, and are able to recognize outward expressions for assent and dissent, they can agree to each others utterances, without having any translation manual at hand, although their ontologies differ. So Quine's argument is that we may differ radically in our beliefs, theories etc., but it is nevertheless possible to find a common ground, viz., agreement on truth and falsity of occasion sentences.

Petra Stoerig provides good justification for this view in her interesting analysis[4] of visual observation processes based on the presence of various types of brain damage and the associated loss of different aspects of one's vision. She describes three steps in the process of visual observation, or seeing. Firstly, we can distinguish a purely *phenomenal encounter* in which an *object* is not yet recognized, but where merely a number of qualities, such as light-dark contrasts and blotches of colour, are perceived. One is aware *that* one is seeing, but not *what* one is seeing. Some people who are born blind has been surgically treated for their blindness so that signals from their eyes enter the brain. But they have had great difficulties learning to interpret visual perceptions, like seeing an object, even where the object is familiar to them. During this period of learning, the patients experience the visual world as a unordered chaos of colours, forms, light and dark, and only after a long time do they learn how to order impressions into figures and objects (many patients fail to do so).

[3] We may carefully observe that Quine claims that there is no fact of the matter which translation is correct, it is not merely that we are uncertain about which is correct.

[4] Stoerig, P. (1996). Varieties of vision: From blind responses to conscious recognition. *Trends in Neurosciences, 19*(9), 401–406.

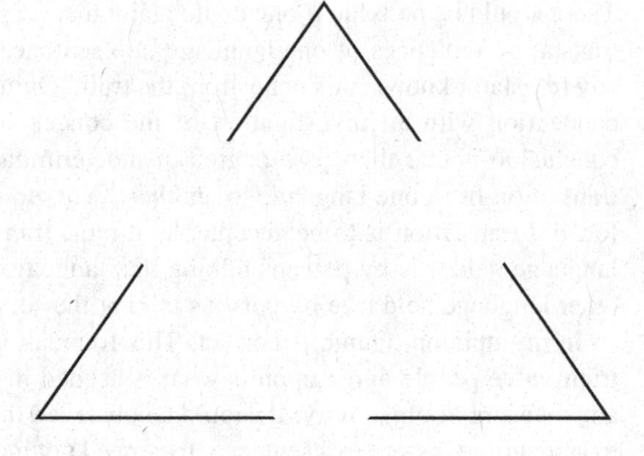

Fig. 4.5 We perceive this figure, without exception, to be a partially hidden triangle, that is, we construct a figure of the given visual elements

The second step in the process of seeing is to *perceive objects* as constituted by separate colours, shapes and contrasts. This results in distinguishing visual objects from their background. However, this perception is still not recognition (Fig. 4.5). There are people who perceive an object, and yet cannot say *what kind* of object it is, even though they have the linguistic comprehension required to do so. For example, one can know what a cat is, and be able to describe what a cat looks like, and yet still be unable to recognize that a certain visual perception of a physically extended object is of a cat. This inability to recognize objects can be found in dementia patients. Oliver Sachs has colourfully described such a patient in his book *The Man who Mistook his Wife for a Hat*. (It regards an elderly man who could no longer recognize his wife, even though he was still able to distinguish various objects in his surroundings).

The third, and last, step in this process is to *recognize* a given visual object as an object of a certain kind. Only when this step is completed does one have a visual observation in the full sense of the word. Hence, the act of perception/observation can be analysed, both conceptually and physically, as a three-step process. When we talk about visual perception – seeing – what we usually mean is the complete process culminating in recognition of an object as an object of a certain kind. Furthermore, it seems reasonable to assume that 'perceiving' by the other senses could also be analysed into stepwise processes, but Stoerig does not discuss this possibility.

The transition from step two to step three in the process of observation given above can be illustrated by the following example. Suppose the object of our perception is a figure that looks like a duck or a hare, depending on how you look at it (see Fig. 4.6). We agree *that* we observe a certain figure, but we do not agree on *what kind* of figure it is. According to the definition of an occasion sentence, in this situation the sentence 'This figure depicts a duck.', is no occasion sentence, if it is the case that some people see the figure as a hare. However, if we reformulate the observational statement so that it is neutral with respect to the kind of object being observed, then it will count as an occasion sentence. In the case of Fig. 4.6, the

Fig. 4.6 Duck or hare?

sentence 'I see a contour that can be interpreted as either a duck or a hare' is a theory-neutral occasion sentence, since everyone would agree with it. The point is that one can formulate observational statements without definite recognition of the object. However, it should be noted that what would elicit agreement from all humans irrespective of culture, and what would do so only within particular cultural contexts, is an empirical question. All that is required for the possibility of theory-neutral observations is the existence of *some* universal agreement among all humans: i.e. that all people have certain features of their cognitive equipment in common.

In medicine, we come across yet another problem regarding testing the effectiveness of medical treatments; namely, the placebo effect. If a patient believes that the medicine he is told to take is effective, then the patient's expectations of being cured are increased. This changes the prognosis of the disease by raising the probability of a positive outcome regardless of whether the medicine has any pharmacological effect! In order to eliminate the placebo effect in the evaluation of treatments, one needs two groups whose members are ignorant as to which group they belong to: a test group and a control group. Furthermore, to minimize the effects of expectations on the data, such as those discussed previously in the timing of mice (expectations being, or resulting from, a sort of primitive theorising) one must make it so that even those administering the treatments are ignorant as to which group is which. This procedure – known as *double blind testing* – is standard for all trials of medical treatments. Since neither the participants nor the people administering the treatment know who is receiving the placebo and who is receiving the effective treatment, the members do not know which outcome to expect. Thus, the purely psychological effects of the treatment can be assumed to be the same in both groups and thus eliminated as a cause of different outcomes. Eventual differences between the test group and the control group can then be attributed to the administered medicine.

It follows from the above discussion that if we want to talk about objective scientific facts, we should restrict ourselves to only those statements that can be expressed by *occasion sentences* in order to limit the extent to which theory colours our judgments. Similarly, we should be careful in saying that a hypothesis/theory is proven, as it is always possible to reinterpret the empirical data hitherto used to support that theory. This possibility is sometimes utilised during scientific revolutions, where one begins to 'see reality' from a wholly new perspective. A classic example is the Copernican revolution.

Before Copernicus, it was generally believed that the Earth lay motionless at the centre of the universe. The thought that the Earth could rotate around its axis was rejected by the following simple observations: If you throw an object straight up into the air, it will fall in the same place from where it was thrown. If the Earth rotated, then the object would fall some distance away from where it was thrown, since the Earth would have moved while the object was in the air. However, this is not the case, as this simple observation is supposed to show. What Galilei did, in his campaign against the traditional view that the Earth is fixed in the centre of the universe, was to interpret this experiment in a different way. He introduced the concept of relative motion and a different definition of the concept of force. Since the days of Aristotle, it was believed that force is required to maintain motion, and that motion ceases in the absence of force. Galilei claimed that motion with constant speed does not require a preserving force. From this it follows that when we throw an object into the air, the object follows the Earth's rotation even when it has left the hand of the thrower. The Earth and the thrown object rotate with the same speed; and so the object falls in the same place from which it was thrown, given that it was thrown vertically.

This example shows that we sometimes discard a theory or hypothesis without making any new observations. As in the previous example, we sometimes 'merely' interpret old observations in a new way. Such reinterpretation is nothing other than a change of one's auxiliary assumptions. In the case of the thrown object above, the essential point is what one means by *force*. Do forces cause motion, or do they cause changes in motion? In the argument against the Earth's rotation, the sentence 'a force is required to maintain motion' is an auxiliary assumption. This view is so natural that one is easily led to claim that one can 'see' that the Earth does not move when an object thrown upward falls precisely from where it was thrown. (The most profound change from Aristotelian to Galilean theory of motion is the conceptual change from the idea 'force causes motion' to the idea 'force causes change of motion'.)

Again, a theory or hypothesis can be strong, believable and very probable, and yet the cautious would refrain from saying that it is verified. Our knowledge is always provisional, more or less.

A sceptic would draw a different conclusion; namely, that we can never know anything with certainty and that all propositions can be doubted. The latter is correct, but one cannot doubt everything simultaneously. For, every time we doubt a proposition, that doubt is based upon another proposition (or more) that we take to be true. Our epistemological standpoint should thus be described using sentences of the form 'I do not believe in *p* because I hold that *q* is true, and *q's* truth supports the belief that *not-p* is true'. It is certainly possible to replace *p* with any statement whatsoever, but this does not imply that one can doubt everything at the same time. Even if there is no secure foundation for the empirical sciences, it is reasonable to assume that most of what we believe is true, while some of the propositions we take to be true are false.

4.5 Observations and History: On Source Criticism

Questions regarding the reliability and theory-dependence of observations are given a different character when applied to history. Students of history do not observe historical events, in the strict sense of the word. However, they do observe the various *remains* (e.g. bones, tools, and ruins) as well as *narrative sources* (e.g. letters, ledgers, and other writings). Only after interpreting and evaluating these historical sources do they discern the historical facts. In turn, these historical facts become the basis of various historical hypotheses.

History as an academic discipline underwent a profound change at the end of nineteenth century, first in Germany. The method of *sourch criticism* was introduced by Leopold von Ranke, (1795–1886), who was professor at Humboldt University in Berlin for a long time. Here is a brief exposition of the main principles.

It is fairly obvious that one cannot accept the contents of letters, reports, or minutes as historical facts without proper analysis. Consider, for example, a letter in which the author describes a historical event. One can pose a number of questions as to the truth of this letter. These questions are usually ordered into three groups; *proximity, bias* and *dependence*.

Proximity The first question regards the sources' proximity to the described event; did the author of the source see the event take place? How much time has passed between the acts of observation and their recording? We all know that the memory is notoriously untrustworthy; when describing an event that took place many years ago, it is doubtful that the resulting description will be accurate. Even worse are those descriptions written by people who did not observe the event themselves, but merely relate another person's observations. We are naturally led to two rules of thumb: (i) if possible, use sources that are contemporary with the events in question, and (ii) that are first-hand. Of course, these two principles can conflict. A contemporary source can be a second-hand source, and a first-hand source can fail to be contemporary. However, experience has shown that if these two principles conflict, then the contemporary second-hand source is usually more reliable than the later primary source.

Bias The next point in the critical evaluation of sources regards the author's intentions with respect to what he is writing. It is possible that the purpose of the text is to impress a particular view upon the reader, rather than give an objective description of the events concerned. A classic example from Sweden's history is Erik Dahlbergh's account of the Swedish army's crossing of 'Det Store Bælt' (The Big Straight) in Denmark, in the year 1658. (Erik Dahlbergh (1625–1703) was a

Swedish soldier, finally field marchal). That year the winter was very cold in Scandinavia, so cold that the Big Straight froze and made it possible for the Swedish army to march towards Copenhagen and conquer Denmark. In his so-called 'diary',[5] Dahlbergh describes the event as if he is one of the central actors of this brave military action. This 'diary' was long considered a reliable document. In modern times, however, this account has been proven false. Dahlbergh's 'diary' is a retrospective construction, written long after the events transpired. It is now seen as an important part of Dahlbergh's campaign to promote his own views and boost his reputation. Critical analyses of this kind are based on an author's personal bias and aim to pinpoint the intentions for writing his or her work. Thus any content that expresses, or is in line with, the author's personal bias is automatically discarded from a historical study, unless that bias is the subject of particular investigation.

Dependence A third form of source criticism regards the possibility that the author was instructed, threatened, or otherwise forced into producing the source. Note that this includes situations in which the author is unconsciously coerced, as in the case of societal conventions. For instance, modern historians do not believe accounts of witchcraft, coitus with the devil, or religious miracles, as these phenomena conflict with general scientific knowledge and can be satisfactorily explained in psychological terms. Another example is found in the Marxist repertoire. Contained in the Marxist perspective is the view that a person's reasons for his actions are secondary effects of objective, material and societal processes. Such a critique is based on the notion that actions of agents are determined by objective historical processes. Thus according to the Marxist perspective, the reasons and arguments an agent gives for her actions or points of view are not relevant, no matter whether she is honest or not. Motives are not historical causes. (Opponents of Marxism were therefore, by definition, deemed representatives of the bourgeoisie, and therefore unable to see the truth. This is a typical *argumentum ad hominem*!) This source criticism thus builds upon the author's dependence on various norms, values, assumptions, historical mechanisms, etc. Furthermore, it is this dependence that makes us doubt the truth of the author's statements.

It is apparent that the utilization of these principles of source criticism in particular cases leaves plenty of room for the historian's personal judgment. Can one then truly say that what is left after a critical source evaluation of some historical material is indeed historical fact? It depends on what one means by 'fact'. If we equate fact with 'considered to be a fact', then it is so by definition. However, if by 'fact' we mean the content of true statements, then it is not the case.

---

[5] An analysis of the paper in the diary shows that it could not have been written earlier than the 1670s, though it was probably written much later. The book draws on actual diaries and almanacs that Dahlbergh authored, but the contents of the so-called diary do not agree with other descriptions of the crossing. See Englund, P. (2000). *Den oövervinnerlige: om den svenska stormaktstiden och en man i dess mitt*. Atlantis, p. 565.

This is in no way unique to the historical sciences. In all empirical sciences one is painfully aware that what is presently considered secure fact can later show itself to be false. Truth exists, but we do not have any guarantees of finding it.

Whether history treats its material as a source or as relics is not a trivial matter. An informative text can be entirely unreliable as regards the events described therein, but it may still be useful as an expression of the author's attitudes, values and beliefs. An example from Thucydides' *The Peloponnesian War* should illustrate this point. Pericles, eulogizing over the fallen soldiers, describes the norms and habits of Athenian citizens:

> Nor are these the only points in which our city is worthy of admiration. We cultivate refinement without extravagance and knowledge without effeminacy; wealth we employ more for use than for show, and place the real disgrace of poverty not in owning to the fact but in declining the struggle against it. Our public men have, besides politics, their private affairs to attend to, and our ordinary citizens, though occupied with the pursuits of industry, are still fair judges of public matters; for, unlike any other nation, regarding him who takes no part in these duties not as unambitious but as useless, we Athenians are able to judge at all events if we cannot originate, and, instead of looking on discussion as a stumbling-block in the way of action, we think it an indispensable preliminary to any wise action at all. Again, in our enterprises we present the singular spectacle of daring and deliberation, each carried to its highest point, and both united in the same persons; although usually decision is the fruit of ignorance, hesitation of reflection. But the palm of courage will surely be adjudged most justly to those, who best know the difference between hardship and pleasure and yet are never tempted to shrink from danger.[6]

If we ask what the text says about Thucydides' view on Athenian life, instead of questioning the accuracy of what Pericles said, or even if it was Pericles who said it, we are not treating the text as an informative source, but rather as an ancient relic that tells us something about how the Athenians saw themselves. Such a treatment can be just as interesting as asking what Pericles actually said.

4.6 Summary

All empirical sciences are tested against data in the form of observation reports. For such testing to meaningful, observation reports must be independent of the theory to be tested. This can be achieved, although observation reports, in the usual sense of the word, often are theory-laden to some extent. It is crucial for the objectivity and trustworthiness of any science that data are acceptable as such both by friends and foes of the particular theory, which is tested by those data.

[6] Thucydides: The History of the Peleponesian War, book 2, chapter 6, (trans: Crawley, R.). http://classics.mit.edu/Thucydides/pelopwar.2.second.html

Further Reading

Ellis, B. (1968). *Basic concepts of measurement*. Cambridge: Cambridge University Press.
Quine, W. V. O. (1990). *Pursuit of truth*. Cambridge, MA: Harvard University Press.
Quine, W. V. O. (1995). *From stimulus to science*. Cambridge, MA: Harvard University Press.
Suppes, P., & Zinnes, J. L. (1963). Basic measurement theory. In R. D. Luce, R. R. Bush, & E. H. Galanter (Eds.), *Handbook of mathematical psychology* (Vol. 1, pp. 3–76). New York: Wiley.
Tosh, J., & Lang, S. (2006). *The pursuit of history: Aims, methods and new directions in the study of modern history* (4th ed.). Harlow: Pearson Education.

Chapter 5
Qualitative Data and Methods

We classify as we can. But we do classify.
Claude Levi-Strauss

5.1 Introduction

The term 'qualitative method' refers to a number of rather different scientific methods within the human and social sciences, such as Hermeneutics, Grounded Theory, Phenomenology and Ethno-methodology. The contrast to 'qualitative' is 'quantitative'; Qualitative methods aim at the collection and analysis of non-quantitative, i.e. qualitative data. But what does this amount to?

As we saw in the previous chapter, if one wants to make a quantitative comparison between collected data, then these data must be such that one can at least compare differences between observations. That is to say, it must be meaningful to ask oneself if the difference between the items o_1 and o_2 is greater, or lesser, than that between o_3 and o_4. There are many cases within the humanities and the social sciences in which it is either impossible, or inadequate, to make such comparisons. For example, take Emile Durkheim's concept of *degree of integration in a society* discussed in Chap. 3. Naturally, one must measure various variables, such as how many parties or holidays are celebrated in a society during a year, in order to get some sort of measure of the degree of integration. However, no single variable will be sufficient for a good picture of the degree of integration. One must somehow sum over all the variables studied such that each addition is based on some more or less arbitrary rule for how the different factors are to be weighed. Hence it is not meaningful to operationalize the concept *degree of integration* to an extent that results in an interval, or quotient scale. One must be satisfied with the ability to make a rank ordering. Another way saying the same thing is to point out that 'the degree of integration' is not a quantitative concept, even though it may appear to be.

Those that use qualitative methods do not usually motivate their choice of method by arguing that quantitative measurements cannot be made. Rather, they argue that the particular phenomena under investigation, and the associated questions one may pose, are qualitative in nature. In these situations, one does not want to find out *how much* of something there is, but rather *what type of character*

L.-G. Johansson, *Philosophy of Science for Scientists*, Springer Undergraduate Texts in Philosophy, DOI 10.1007/978-3-319-26551-3_5

something has. For example, a hermeneutic would perhaps want to interpret the content of a historical or literary text, while a phenomenologist would want to research how certain individuals experience their existence, their life and the world around them.

To say of something that it is thus and so is to decide which concepts the particular phenomenon falls under. To make such a decision is to make a *classification*. We can thus draw the conclusion that qualitative methods can be characterized by the fact that the data being collected are classified; that is, qualitative observations form a nominal scale. But this is not sufficient for a definition of the qualitative method, since this characterization is too broad. By this definition, the determination of blood types would count as a qualitative method. In one sense, it is obvious that the determination of someone's blood type is a measurement that results in a qualitative statement. However, the term 'qualitative methods', as it is normally used, means something different.

It strikes me that all qualitative methods have yet another thing in common; namely, that the phenomena one is studying are *meaningful*, and the primary aim of qualitative methods is to identify the *meaning* or *significance* of texts, symbols, actions, and so forth. Meaning and significance are concepts that belong to the *intentional realm*. Therefore, I propose the following definition of qualitative method:

Definition of qualitative method: A scientific method is qualitative if and only if it aims at the classification of phenomena with respect to categories containing an explicit, or implicit, intentional component.

But what is meant by 'meaningful', 'significance' and 'intentional'?

5.2 Intentionality and Meaning

The word 'intention' is often treated as synonymous with 'purpose' or 'aim'. Intentions are involved in human actions. However, within philosophy 'intention' has been given a broader meaning. The philosopher Franz Brentano (1838–1917) observed that all descriptions of mental phenomena can be analysed into two components: the mental *act* and the *content*. When Lisa says 'I hope we will have a nice summer', the mental act is the act of hoping, and it has the content 'we will have a nice summer'. One cannot hope for nothing in particular. One must hope for something definite; that is, the act of hoping must have content. This content, towards which the act is directed, is often called the act's *intentional object*. Brentano held that all mental phenomena are thus directed towards an intentional object.

Most, or perhaps all, mental states fall into the following three groups:

- perceptions,
- emotions,

- propositional attitudes (believing, hoping, wishing, thinking, etc.).

All of these mental acts are in a certain sense *directed* towards an object. When we observe we direct our attention to some object, which becomes the *intentional object*. (There is an important distinction to keep in mind here: sensations are not, while perceptions are, object-directed. Pain is a sensation not a perception, since there is no object towards which the pain is directed (headache is not pain directed towards the head, it is a type of pain), while seeing e.g. a fox is a perception.) Our emotions are also directed: we are mad *at* someone, thankful *for* the presents, and anxious *over* the impending examination. In all these situations, we indicate the intentional component of the mental phenomenon with a preposition followed by a description of the object of intention, towards which the mental act is directed. Finally, propositional attitudes such as believing, hoping, knowing and wishing are directed. The object toward which an attitude is directed is more often called its 'content' and is expressed by the proposition following the word 'that'. Such propositions may even contain objects of fantasy such as Santa Claus or Hamlet. (There is a philosophical problem with the semantics of sentences where such objects are non-existent; but this is beyond the scope of the present discussion.)

Propositional attitudes are closely connected to actions; usually we call something done 'an action' when we take for granted that it is connected to a belief or conscious desire. When we think that the doing is not associated with any belief or conscious desire we call it behaviour. Hence, we should distinguish between action and behaviour: actions always have an intentional component, whereas behaviours don't.

Choosing to do nothing in a given situation with the intent that some event will take its course without interference can also be considered an act precisely because the agent has a belief about what will happen, and a desire that the expectation materialize.

We can thus describe all mental phenomena by indicating the mental act being performed and the object towards one's mind is directed. Directedness, or intentionality, is a necessary component of mental acts.

Meaningful phenomena are phenomena that either contain an intentional component themselves, or else are more indirectly constituted by an intentional property. Thus actions, and the results of those actions, are meaningful phenomena in the same way as texts, social institutions and conventions, values, norms and artefacts. This is because all of the above phenomena are either directly constituted by intentions (collective intentions constitute social institutions, as will be discussed in Sects. 5.8 and 5.9), or the result of such intentions.

How is one to research the meanings – the intentional components – of these phenomena? Can they be observed? Do they exist per se, or are they created in the very act of observation? I shall attempt to illuminate these questions in the following three examples.

1. A stone found during a stroll through the woods seems, at least at first, to have no meaning. Yet, if we look closer, we may find that it is not just any old stone, but that it shows signs of having been formed by human hands. This unusual shape

gives the stone a sort of meaning. Though we may not be able to give details as to what that meaning is, we can observe that it has meaning in that its form is the result of conscious human action. Once we have convinced ourselves that this form is human-made, we ask ourselves, what is this meaning; why does the stone have this particular shape? Is the stone a tool or a cultural expression? In asking such questions we assume that someone had a particular purpose in mind when he, or she, crafted the stone. It is an archaeologist's job to interpret the stone's features in order to discover what this purpose was; thus allowing us to understand how the stone was used (e.g. to clean animal hides).

Does this stone have meaning that we discover, or is it that the stone only acquires a meaning once it has been studied? Difficulties arise here from common language use as well as ontological assumptions. If we say that the stone has meaning that we can discover, then this meaning is a property of the object, which is independent of the observer. If, instead, we say that the stone acquires a meaning *for us*, then this meaning is more likely a relation between the observer and the stone. One cannot then say that the stone has any meaning in itself, but rather that meaning arises in the observer's interaction with it. In this case, I think it reasonable to say that the stone has meaning; namely, its function with respect to the way humans used it. We can find, or fail to find, this meaning, but we should note that the stone's meaning is not a physical property of the stone like its mass or shape, but rather an intentional one.

2. The meaning of certain actions and artefacts is a much more difficult question. Money is an everyday example of this difficulty. Certain physical objects such as thin cylindrical pieces of metal, small printed notes, and electronic currency are considered to be money. However, the monetary value of these objects is not a physical property of them. Of course, a 10 £ banknote must have certain distinctive physical features, such as the print of a 10 £ pound note. Yet, it is not the print that gives the money its value. Rather, it is a *collective agreement*, or attitude, that constitutes its meaning. When a country suffers inflation, this collective attitude changes dangerously as people loose faith in their currency. Even though no physical aspects of the monetary objects have been changed, the normal value still goes down. This shows that what changes is, in fact, the collective attitude.

Most of the time, these collective attitudes are unconscious. We do not normally think of banknotes as pieces of paper that have come to be treated as money by way of a collective agreement. It is hardly ever the case that someone thinks, 'I am handing over a piece of paper, and I recognize that everyone considers this piece of paper to be money'. One cannot look at a banknote, or an electronic currency statement, and see what makes it money. In order to determine what makes certain objects money we must observe how people interact with these physical objects, or physical states (electronic currency is a physical state of a bank's computer). For example, it was through direct observation of Bosnia and Croatia during the war from 1991 to 1996 that we could see how the Deutsche Mark and cigarettes were treated as money, but the official Yugoslavian Dinar was not. The Dinar had lost all its value.

Assuming that collective intentions, on which social institutions are founded, are fictitious creations of a researcher describing these social institutions, we are faced with the possibility of not being able explain a large number of phenomena and human behaviour. From a methodological point of view, these collective intentions play a role similar to that of electrons in physics. We cannot *directly* observe electrons[1] or intentions; yet, we have good reasons to believe that both exist. If we were to assume that, for example, electrons do not exist, then a large number of observable phenomena would be nearly impossible to explain or understand. Similarly, it is rational to assume the existence of collective intentions as the constitutive elements of social institutions, since this assumption allows us to understand many aspects of the human social behaviour. Thus it is reasonable to say that the meaning that social institutions have, i.e. the collective intentions behind them, both exist and their effects can be observed. Importantly, the institutions so spawned are not the result of a single individual's intentionality, but rather their meaning is shaped by the interaction of many. It is this *collective* aspect that ensures objectivity and observer-independence of these kinds of social institutions. In summary, we can recognize a particular social institution by observing how people relate to each other and the physical objects with which they interact.

3. Finally, there is a class of phenomena for which there are three very different positions regarding their meaning. Examples of this class notably include works of art. The meaning of a work of art can be (i) the intention of its creator, (ii) some sort of inherent quality of the artefact itself, or (iii) the resonance the artefact excites in the present culture.

Thus, the question regarding a work of art's true meaning is ambiguous. In particular, this is the case within the literary sciences, where one finds an on-going discussion regarding whether inquiry into the interpretation of a literary work should be directed at the author's life and context, the text itself, or the work's interaction with its present social context. This question strikes me as regarding what is most interesting, or fruitful, and not as a question about what is most accurate and true.

Meaning is a type of relational property between certain physical objects and states, a relation between one or more people and an object, state or event. However, the researcher investigating these phenomena does not shape their meaning. Meanings are objectively existing phenomena in so far as they are properties of social reality that exist independently of what the researcher, or any other *single* person, thinks; they are collective phenomena. So statements about these phenomena are true or false, independently of what the researcher, or any other individual believes; thus, such *statements* are objective. Searle[2] coined this *epistemic*

[1] Physicists nowadays claim that they observe electrons, but obviously they mean that they observe electrons using highly advanced technology based on quantum mechanics. So the 'observation' is highly indirect.

[2] John Searle, (1932) American philosopher, professor at Berkeley.

objectivity. Another form of objectivity observed by Searle is the ontic; some *things* exist in subject's minds, others exist in the world and would continue to do so were there no minds at all. The former are *ontologically subjective*, the latter *ontologically objective*.

Statements about social meaning are epistemically objective. But social meaning could not exist without the existence of individual minds having intentions and beliefs. Should we then draw the conclusion that social meaning is ontologically subjective? Searle's answer is yes.[3]

Since propositions concerning these phenomena are objectively true or false, it is possible, in principle, to analyse them, and thus make objective statements about them. I write 'in principle' because it is sometimes impossible to know what someone who has long since been dead intended in regards to some action. Yet, it is clear that this person had *some* intention in regards to his or her action; hence, propositions concerning this intention are objectively true or false regardless of whether we can ever know them.

This conclusion is not generally accepted. From the fact that all the phenomena above are social constructions, some draw the conclusion that objective propositions regarding these phenomena are not possible. I shall discuss this counterargument and the various interpretations of social constructions in Sect. 5.9.

Given this discussion of intentionality, it is now appropriate to investigate a couple of examples of qualitative methods that exemplify the general discussion surrounding intentionality and qualitative methods. These methods are hermeneutics and grounded theory.

5.3 Hermeneutics

Hermeneutics is often characterized as a tradition of research aimed at how one reaches an understanding of various phenomena, such as texts or historical events. According to a number of hermeneuticians, understanding requires empathy, the ability to place oneself in another person's situation. This is the central thesis of R. G. Collingwood,[4] see his (1994).

We often use the word 'understanding' in a different sense. We can, for example, ask someone if he understands Relativity Theory. In such a case, we cannot attain understanding through some hermeneutical process, but rather through describing the logical connections between concepts in the theory and how it is to be applied to concrete physical situations. Understanding in the hermeneutical sense is something else, viz., the understanding of the *meaning* of something that is bestowed with an intention, a human, a group of humans, or the product of human actions. To

---

[3] See Searle (1995), p. 12.

[4] Robert George Collingwood (1889–1943) British philosopher, historian and archeologist.

understand a human is thus to understand what he or she wants to express or achieve through his or her various actions, writings, or art.

Understanding in the second sense is attained through *interpretation*. Hermeneutics rightly stress that no interpretive process starts 'from scratch'. Rather, we always approach the unknown with some *preconceptions*. These may concern what counts as an explanation, how people typically think and act, and what linguistic or conceptual frameworks are applicable in the specific case. From these preconceptions we make a preliminary, and often crude, interpretation that forms a starting point for further studies. These further, more detailed, investigations can either strengthen our preliminary interpretation, or lead to their revision. Thus in the process of interpretation, one begins with an initial conception of the whole and then uses this to inform a more detailed study of the parts. This detailed study of the parts then serves to inform a reassessment of the whole, which then serves to inform a re-evaluation and deepening of the study of the parts, and so on. This shifting of focus from interpreting the whole to studying the various parts and back again is referred to as a hermeneutic circle. This circularity ceases when the process of interpretation attains a sufficient degree of coherence between the interpretation of the parts and the whole. (There are other formulations of the goal of a hermeneutic process, but I shall not discuss them here.)

Thus, the central activity of hermeneutical research is interpretation. But what is the link between interpretation and classification?

Firstly, the practicing hermeneutic must determine a limit as to the kinds of phenomena that have meaning. This is, in effect, a classification. Sometimes there are interesting borderline cases; one such is so-called hysterical diseases. Paralysis of the arm or leg without any apparent physiological cause was interpreted by Sigmund Freud[5] as a hysterical disorder. Freud considered this kind of illness as an unconscious reaction to conflicts with respect to some unbearable social situation. Since his patients were predominately upper class women, he believed these symptoms to be manifestations of an unconscious plea for attention. In 1900, upper class women in Vienna, where Freud lived, were constrained to a fixed way of life. Their only purpose consisted in producing heirs for their husbands and acting as decoration in social situations, which did not leave much room for personal goals and requirements. Thus, Freud heard a cry for help: care about me, I am important, I need to be taken seriously! A hysterical disease was thus a means to gain attention and, given that Freud treated such cases as phenomena with meaning, required interpretation. According to modern physiological studies, however, hysterical diseases are actually injuries in the motor-skill centre of the brain, which controls the function of the extremities. This would appear to refute Freud's theory, but is that true? Could it be that the very social situations in which the afflicted women found themselves were the causes of their brain injuries? This dispute touches upon the fundamental question as to whether or not these diseases have

[5] Sigmund Freud (1856–1939) was an Austrian doctor, first active in Vienna, later in his life in London, who started psychoanalysis.

meaning; and thus also upon the question as to whether or not one should regard psychology as a human or natural science.

In my opinion, it seems that the question regarding whether to describe these states in terms of mental or physiological terms is purely a matter of appropriateness. Once we know more about such phenomena we will, I think, come to understand that they are merely two different ways of describing the same thing; mental phenomena are physical phenomena, although described in psychological terms. More about this in Chap. 12.

The first classification the hermeneutic makes is thus the determination of what can be the object of interpretation. However, the eventual process of interpretation is itself a sequence of classifications. Let us examine an example from Swedish history to illustrate this point. The document that established the so-called Kalmar union between Sweden, Denmark and Norway, written on July 20, 1397, contains a number of irregularities. First, it is not written on parchment, but on paper, which was highly unusual for important documents of the time. Furthermore, several seals of important persons are missing. Interestingly enough, it was only 7 days earlier, on July 13, 1397, that these persons had signed and attached their seals to the so-called coronation letter in favour of king Erik of Pomerania. Given these anomalies, the hermeneutic historian asks the following: are they historically significant, or are they purely circumstantial? Furthermore, if these anomalies are significant, then what do they signify? Should we interpret them as the silent protest of certain groups in Sweden and Norway against queen Margareta's claims to the throne (Erik was under-age after all), or as something else? There are a number of further classifications that remain to be made.

An important problem in connection with the hermeneutical method is the choice of criteria for accurate interpretation, which I shall discuss more generally in Sect. 5.8.

Without delving deeper into a discussion about hermeneutical methods, we can conclude that such methods involve classifications of meaningful material with respect to intentional categories.

5.4 Grounded Theory

Grounded theory is a tradition within sociology inspired by Symbolic Interactionism, whose ontological thesis is that social interaction is an exchange of meaningful symbols. In Grounded Theory, this thesis is applied to a strong, empirically oriented method. The starting point is a critique of the classical sociologists' (Marx, Weber, Durkheim) 'grand stories' for being too abstract and with little connection to empirical data. Glaser and Strauss (1967), the seminal work on Grounded theory, claim that any new theory should in contrast start from bottom and be generated on the grounds of empirical data. Once these data has been collected, they are treated using a method called *coding*, which results in a set of *categories*. The method could be described as consisting of four steps[6]:

[6] This description is adapted from Alvesson and Sköldberg (2008).

- Read the text (field notes, interviews or documents) word for word, row for row, or at least paragraph for paragraph.
- Always ask yourself, under which everyday, or otherwise common sense, categories does the data in the text fall?
- Continuously note down these categories and further data that fall under them.
- Inspect the data and categories in order to find possible further enriching properties.

This collection of bullet points may seem a bit vague to the outside observer, for one is compelled to ask what is meant by 'data', 'categories', and the properties thereof. It is obvious that we need concrete examples of this method in order get a clear picture of how it is supposed to work. However, one can perhaps make the following reflections without such examples.

1. There is a clear inductive attitude in the method, and it seems taken for granted that classification of data into categories arises naturally and without the aid of theories.
2. The method draws on Bacon's inductivism, in so far as classification is said to arise naturally via a detailed study of the empirical data. Bacon's inductivism has been criticized for theoretical naivety; the sorting of objects and phenomena into categories is at least partly an active process of the mind. This critique also pertains to Grounded Theory.
3. If one must use only those categories that are intelligible in advance, one runs the risk that research being able to produce trivial knowledge only, since new theories require new conceptualization.

Alvesson and Sköldberg (2008) share similar sentiments. After a discussion of an example of coding they write:

> The coding will, despite good intentions, not portray 'reality' unequivocally through an objective, unambiguous, secure and rational procedure. On the contrary, the researcher will, in light of his/her unconscious frames of reference, interpret what he/she sees in front of him/her. Two problems with an unconscious frame of reference is that one advocates that reference frame in the form of the bias that arises from common-sense classification, and that one places far too much energy on detailed coding.[7]

This is not the place for a critical investigation of the pros and cons of Grounded Theory, but one thing is abundantly clear. That is, a central point of this qualitative method is the classification of empirical data. Since Grounded Theory is a method for understanding the meaning that people give to various types of phenomena, this classification is done with respect to intentional categories.

Without any further discussion of other qualitative methods, I am prepared to make the generalization that my definition of them is correct; namely, all the methods that are normally called qualitative are captured by two defining criteria: they are classifications, and that the resulting categories are intentional.

[7] op.cit p. 84.

5.5 The Intentionality of Observations

According to certain methods of research, e.g. Grounded Theory and Ethnomethodology, one is supposed to passively observe the object of research and then theorize about what one has experienced. As a general methodology, this is naïve. Observation is an active mental process. It is intentional, directed towards certain features of the perceptual field, as well as some features that are more or less unconscious. With every blink of the eye, our sense organs are flooded with impulses, and the brain's first task is to sort out most of them so that we can focus our attention. (There are tragic examples of people that, more or less, lack this selectiveness; they register all impressions but are unable to sort them regarding salience and to use them for any purpose.) The question arises, 'Of all the impressions our sense organs take in, what is sorted out, and on what grounds is this done? How do our intentions govern our perceptions? All we presently know about the answer to these questions is that it can probably be given a general description in evolutionary terms. The ability to direct one's attention toward features in the environment relevant for survival and reproduction is a consequence of natural selection. Of course, we humans often direct our attention towards things other than those required for survival, but the important point is that at some time we developed a nervous system that has the ability of focusing our attention on a great variety of things. It is correct that observation is governed by the observer's cultural frames of reference, but it is based on this fundamentally biological aspect of observation, viz., that it is selective and goal-directed.

This gives strong empirical evidence for the assumption that we are unable to observe without any presupposed concepts, or rather, observation by its nature is application of concepts.[8] This does not mean that one is in possession of articulate hypotheses about possible observations, nor that we should not endeavour to rid ourselves of prejudice and expectation to the maximum extent possible. Furthermore, it does not mean that initial data has no role to play in guiding test hypothesis choice; initial preconceptions and test-hypotheses, though related, are distinct entities.

In the previous chapter observation was analysed as a three-step process; (i) to acquire phenomenal experiences; (ii) to synthesize these experiences into an object; and (iii) to recognize this object as an object of a specific kind. The second step in this process introduces the directedness of the act of perception, that perception by its nature is object-directed. This directedness encompasses the ability to focus one's attention on an object in the field of perception, and to distinguish it from the rest.

The object-directedness of perception is rather obvious when we think of visual perception, but perhaps less so when we consider our other senses. However, this is

[8] This conclusion is similar to Kant's view, expressed as 'intuition without concepts is blind'. (*Critique of Pure Reason*, A51, B75).

mainly due to the fact that we naturally think of visual objects, bodies and pictures, when we read the word ‘object’. If we generalize and remember that the word ‘object’ as used in philosophical discourse means whatever can be talked about and thus subject to predication, we may appreciate that for example hearing also is object-directed. We hear objects such as other persons’ voices, musical themes, and all kinds of recognizable sounds. In addition, we are all acquainted with the phenomenon that we might hear something without listening to it. This is the difference between having a phenomenal experience of sound and having an audial perception.

In the previous discussion of Grounded Theory, an instruction was quoted advocating the reading of a text word-by-word, line-by-line, and so forth. This attention to detail may seem trivial, but it is not the physical aspects of the text one should focus on. Rather, one should focus on the text in a way that regards it as a linguistic entity with syntax and semantics, as well as the pragmatic aspects of the author’s intentions, when they are relevant. However, how does one know if they are relevant? Again, we need a preliminary theory, which can be as simple as a gut feeling, or a purely physiologically based preconception inherent in the conscious act of perception. Thus, observation is not entirely unbiased.

In summary, the process of observation is the transformation of a collection of qualitative aspects of experience into objects. These objects can often be recognized as objects of a specific type (cats, cars, events, etc.), and once we have reached this third step in the observation process, we are mature enough to formulate any number of observational statements.

The central philosophical question in this discussion is whether one can say that the perception of an object is entirely theory-neutral or not. Personally, I am inclined to answer that the perception of a physical object is, in most cases, the same process for all people regardless of cultural or theoretical predispositions. The main argument is based on observations of language learning, both in the case of the children’s learning its native language and of anthropologist’s learning the language of native cultures. An extreme example is learning the language of some tribe in the Amazonas or in Papua New Guinea. Some of these tribes still live as gathers-hunters with almost no contact with western civilisation. Despite great cultural differences and without any previous knowledge about their beliefs, anthropologists are able to learn their language and to communicate with them. How could that process begin? It seems utterly plausible that one must do the same as when one learn small children the first words; one begins with words for physical objects in the immediate surrounding, point to them and say the word and encourage by gestures the other to say something in his language. If you want to know the natives word for fish, you may point to some fishes and say ‘fish’ and encourage the other to say a word in his language. Perhaps he has no general term corresponding to ‘fish’ but only for different species of fish; but that will sooner or later be discovered by the anthropologist. This would not work if not both parties discerned the same physical objects in their surroundings.[9]

In the previous chapter I suggested that if there is doubt about whether a given observation has been made, it is always possible to make use of more primitive formulations in the verbal expression of that observation. It is always possible to derive a sufficiently primitive observational statement such that two researchers can agree to it, no matter how much their theoretical backgrounds differ. This applies just as well to qualitative research. For, in many cases it is easy to agree as to the words contained in a text, or – for that matter – what the text says, even if one disagrees about how this information should be interpreted. There are hardly any theory-independent observations, in an absolute sense, if by the word 'observation' we mean the aforementioned three-step process. However, theoretical disagreement at the third level does not entail disagreement at the lower level of observation.

When we study raw data, or make observations, we formulate observational statements. Every such statement is a declarative sentence. Declarative sentences are comprised of an individual term and a predicate (see Appendix). The individual term refers to the observed object, which can be a physical object, a person, an act, a smell or a text, anything that can be identified and re-identified as *the same thing*. The predicate, on the other hand, is a general term that describes a property that several objects may have, or a relation between objects. The result of an observation is thus the claim that the observed object satisfies a certain predicate. This corresponds to the third step, the recognition of an object. The application of a predicate to an object is thus a classification, since one is sorting that object into the first of the two categories (i) those that satisfies the predicate, and (ii) those that do not. In other words, observational statements are the result of classifications. The interesting question is now: Are these classifications correct, informative, interesting or useful?

The French cultural anthropologist Claude Lévi-Strauss (1908–2009) observed that all humans, scientists and 'primitive peoples' alike, classify. The difference between the two is the choice of categories. The quotation from Lévi-Strauss at the beginning of this chapter, 'We classify as we can. But we do classify', expresses precisely this point. The difference between 'modern' scientific thinkers and so called primitive people is not that the former make classifications of phenomena that the latter are unable to make, but that one makes classifications according to different criteria. Doing science is to categorize phenomena in a more useful and considered way.

Are the scientific classifications more correct than those made by Amazonian natives (Lévi-Strauss' specific field of research)? Classifications of plague and volcanic eruptions as expressions of the will of the Gods seem 'primitive' to us and thus incorrect. But is it not likely that 'primitive' peoples would reject our

[9] Quine argued in his (1960) that one cannot determine which ontology the native has only by agreeing on translation of complete sentences. So the native may not after all, view the visible world as inhabited by physical objects. His argument is in my view convincing as far as it goes, but there are other reasons to think that all humans cognize the visible world as inhabited among other things, by physical objects.

scientific explanations of plague as a disease caused by transmittable microorganisms, or volcanic eruptions as caused by movement of the tectonic plates, as irrational and much more speculative than assuming gods. Is there any objective criterion for ranking these two archetypical explanations?

The answer is that the best theory is the one that helps us predict the occurrence of the phenomena in question, thus enabling us to take preventative action. We want to prevent plagues and disasters, or at least be able to reliably predict when they will occur, a goal on which most people, of all cultures, could agree. If we take measures to prevent some disaster using a false theory, there is a chance that doing so will make the situation even worse. A tragic example of this is the 'black death' of the fourteenth century, when people would seek shelter in churches and beg God for mercy. In the crowded quarters of a church, the plague spread more easily.

Modern science has shown itself to be superior in this way, to other modes of thinking in many respects. However, this does not mean that all scientific theories are correct, only that science in general is on the right path. To make scientific classifications, in contrast to primitive ones, means making better predictions, but is no guarantee for truth.

5.6 Are Quantitative Methods Better than Qualitative?

It is sometimes claimed that mature science is quantitative, and thus the proper model for all scientific endeavours. It seems doubtful, however, whether such a claim could be defended. It is certainly a big step forward to have made one's concepts precise enough so that quantitative comparisons are possible. A simple example is the substitution of the quantitative concept '...higher temperature than' for the qualitative concept '...hotter than'. However, is this precision always desirable? Take, for example, a concept like *degree of democracy*. It is certainly of relevance in political science. Can such a concept as *democracy* be reasonably quantified in an uncontroversial way? I do not think so. Democracy contains many different factors, such as fair elections between various candidates, freedom of speech, protection of minorities, etc. A few of these factors may allow for quantification, but how shall they be weighed against each other? Can a low quantity of one factor be compensated by a high quantity of another? Perhaps it is more reasonable to require a certain minimal standard for all those factors that are necessary for democracy. Yet, all this allows us to do is to categorise systems of government into the democratic and non-democratic categories, or at best perform a ranking. It does not allow us to compare degrees of democracy in a quantitative scale.

It seems reasonable to say that *degree of democracy* is not a quantity, but rather a multi-dimensional qualitative measure based primarily on judgment. Assigning numerals to the relevant categories and then deciding the relative weights of each category may quantify this judgment. However, there is no objective basis for how these relative weights are chosen.

The conclusion is that it is not reasonable to say that all scientific disciplines should strive for quantitative methods. The use of qualitative methods does not make for a lesser or inferior science. In fact, there a many interesting scientific questions that are discussed on the basis of qualitative data. This conclusion should not, however, be taken as an argument for ridding certain sciences of their quantitative methods. For every qualitative concept, one ought to ask: Is there an objective way of replacing the concept in question with a quantitative one? Is there a point in doing so? During certain periods the debate in such disciplines as sociology and pedagogy has revolved around this question. Many so-called positivists have claimed that a requirement for proper science is that one must be able to quantify observations in an objective way, whereas critics have claimed that attempts to quantify what is essentially qualitative in nature leads to a loss of the core of this data; namely, its intentional aspects. According to my grasp of the situation, one cannot take a general stance regarding this question; rather, in every concrete situation one must ask oneself what kind of question – quantitative or qualitative – one is trying to answer. One must first have put forth a hypothesis before one can decide what kind of investigation is to be done. Thus, the debate between positivists and anti-positivists should be reformulated as a debate about which questions and hypotheses are we interested in answering.

The positivist would then claim that qualitative hypotheses do not allow for objective testing, since all qualitative data is based on judgments, which are subjective in nature. Therefore, according to the positivist, only quantitative questions are worth researching. The anti-positivist would beg to differ. Hence, the central issue in the debate between the two is objectivity.

5.7 Objectivity and the Use of Qualitative Methods

There is a general reluctance to apply qualitative methods because of their ostensible uncertainty. This uncertainty has nothing to do with classification as such, but with its intentional aspects. No one has ever argued against the determination of blood type because it is a qualitative method (given that qualitative method is interpreted as plain classification), since the method of testing is extremely reliable. The uncertainty of qualitative methods lies not in that they classify using nominal or ordinal scales; rather it lies in the search for the correct *meaning* of the phenomena in question. A condition for claiming that qualitative statements are objectively true or false is that the object involved exists in reality, and does so independently of observation. Thus we have come to the central ontological question: Does this meaning, which is the goal of all qualitative research, have an independent existence?

If one claims that all that exists are physical things in space and time, then it immediately follows that the phenomenon in question (the meaning of something) does not exist, since it is supposed to be a non-physical entity. However, this premise expresses an extreme point of view, and few share it. Even a strict physicalist like Quine accepts that there are non-physical things such as numbers, sets and other mathematical entities. But he would not agree that there are entities that are defined in intentional terms, since he thinks that we have no clear criteria of identity for such things. There are multiple arguments for this point of view, such as the following. In order to claim that a certain abstract entity (e.g. the meaning of a sentence) exists, one must be able to distinguish between the entity and its description or name. It follows that one must be able to determine whether two different descriptions describe the *same* entity, or different entities. To do this, one must have criteria for the identity of such entities. Quine's view is that it is impossible to formulate criteria for the identity of intentional entities, and he draws the conclusion that such concepts should be kept out of science proper. This view is highly controversial, and few have fully supported it, even though the requirement of identity criteria is reasonable.

As I have previously claimed, I think that meanings exist objectively, and can be researched in the same manner as physical things. Yet, this claim was made without offering an explicit example of the, admittedly indispensible, criteria for the identity of intentional entities. Unfortunately, such a discussion would be well beyond the scope of this book.

Objectivity in the humanities and social sciences has been debated from other points of view. Some have claimed that social phenomena are *social constructions*, and that these do not objectively exist. The conclusion that many draw is that the requirement of objectivity should therefore be abandoned. John Searle attacks this view in his book *The Construction of Social Reality*. Searle introduces the concept of *social facts* to emphasize the external and objective character of social phenomena. These concepts, *social construction* and *social fact*, rightly deserve their own section.

5.8 Searle on Brute and Social Facts

Searle's most important distinction, in his (1995), is between *brute facts* and *social facts*. A brute fact is a fact whose existence does not depend on human thoughts, attitudes or knowledge, whereas the existence of social facts does depend on these things. An example of a brute fact is the fact that Mount Everest is 8,848 m high. Notice that it is not the statement itself but its content that constitutes the fact. This brute fact – the height of Mount Everest – does not depend on whether or not humans have measured it, nor does it depend on which sentence we choose for its expression; for instance the same fact can be expressed as 'Mount Everest is 29,493 feet high'. This is a different sentence; but it states the *same* fact. Of course, we cannot *speak* about this fact without applying certain concepts and having knowledge, yet the fact would hold even if no human had ever existed. The statement is true (or perhaps false, if we have measured incorrectly), irrespective of our knowledge about the height of Mount Everest. Thus, that Mount Everest is 8,848 m high is a brute fact.

Certain other facts, on the other hand, do essentially depend upon people's perceptions. Suppose that you are watching a football[10] match on television. It is assumed that you believe it a fact that what you are watching is a football match that is either being played now, or has been played in the past in the case that the feed is not live. This football match is obviously dependent upon the existence of humans. However, it is not enough that there exist 22 players and a ball. There must exist referees, linesmen, spectators and many others that must *agree* that there is a football match going on in order for it to be a football match. If enough people think that it is not a football match, then it is not a football match, even though the movements of the players are precisely the same. The remarkable thing about this type of fact is that it seems to depend wholly on what the people involved *think* about it. Searle introduced the concept of a *collective intention* in order to draw attention to this joint comprehension. If many people are focused on the same object, activity, or event, then we can talk about their *collective intention*.

Searle calls facts that are constituted by collective intentions *social facts*. Of these social facts, the most interesting subcategory is that of *institutional facts*.[11] Searle's prime example of an institutional fact is the fact that certain objects and states of affairs are considered to be money, as discussed in Sect. 5.2. If there were no collective intentions to treat certain objects as money, then these objects would not be money.

The structure of institutional facts is thus that something, X, counts as something else, Y, under certain circumstances: This is a constitutive rule,[12] which we can use as an implicit definition of institutional facts:

X counts as Y in contexts C.

Where the expression 'counts as' expresses a collective intention; the collective intention imposes a certain status, Y, on things previously identified as X. It is this imposition of status that creates the institutional fact.

The basis for such constructions could also in turn be institutional facts. One immediately asks if there must be an ultimate basis, a bottom of brute facts that are not so constructed. Searle's answer is yes, there must be some objects that are not constructed; brute facts are ontologically prior to institutional facts. Institutional facts ultimately depend on brute facts.

Other examples of social facts are 'Kim Philby was a Soviet spy', 'Sweden is a member of the E.U.', and 'John Doe owns shares of Google stock'.

[10] Like all Europeans, by 'football' I do not refer to american football, but to soccer.

[11] According to Searle, examples of social facts that are not institutional facts are collective actions that do not require use of language; for example, the coordinated hunting of wolves'.

[12] Searle uses here the distinction between regulative and constitutive rules. A regulative rule regulates activities, which occur independently of the rule; for example, traffic rules. Constitutive rules, by contrast, constitute the activity, which comes into existence by the implementation of these rules. The rules of chess are a good example; if you are not moving the pieces mainly according to the rules, you are not playing chess.

It is an institutional fact that Sweden is a member of the E.U., since Sweden's membership depends on the attitudes of many people and these attitudes confer a certain status to the kingdom of Sweden. Assume that all humans in Europe and the E.U. suddenly decided that Sweden is not a member of the E.U. Sweden's representatives would not participate in the E.U.'s meetings, customs would not apply E.U. regulations as regards the transport of goods between Sweden and E.U. countries and the E.U. countries would not ask the Swedish government's opinion on internal E.U. matters. In short, everyone would act as if Sweden was not part of the E.U. An expert in international law would perhaps say that Sweden still has the proper documentation stating that it is a member of the E.U., but recall that the premise was that everyone believes that Sweden is no longer in the E.U., even legal experts. That is to say, if no one in the E.U. recognizes these documents, then in all practicality, Sweden is not a member of the E.U. On the other hand, it makes no difference if all documents regarding Sweden's membership are destroyed. For its membership is a social and institutional fact; it consists of the opinions and actions of a large number of people. If all of these people, or the relevant part of them, were to change their minds, then this social and institutional fact would also change.[13]

Would the fact remain unchanged if only some of the people involved, but not all, changed their minds regarding Sweden's membership in the E.U.? Suppose that Sweden's government was of the opinion that Sweden was still an E.U. member, but the French government was not. In such a case, it seems that the matter would be decided in meetings between the heads of both governments, and if not, then at the E.U. court. Sooner or later, the matter would be cleared up.

This discussion concerns the question whether or not a certain fact is the case; that is, how one should characterize the relevant group of people. As far as I can see, there is no general answer to this question, since the context determines what counts as the relevant group. The discussion does not concern the question whether or not an alleged fact is social or brute, but whether it is a fact at all. It is not uncommon for one to be unsure as to whether or not an alleged social fact obtains. This uncertainty applies to brute facts as well. After all, is Mount Everest actually 8,848 m high? Maybe its height was inaccurately measured.

Human beliefs are subjective; therefore one is tempted to say that the social fact that Sweden is an E.U. member is also subjective. However, this conclusion is incorrect. A social fact is essentially a state of affairs that many people consider to be the case and act in accordance with. What many people are in agreement over is objective in a certain sense, namely, *intersubjective*. One must remember that 'objectivity' can have more than one sense, and that one of the most common and important is intersubjectivity.

[13] The written constitution of Soviet Union fulfilled high demands on democracy, but in practice the secretary-general of the communist party was dictator and all kinds of opposition to the communist system was promptly and severely punished. The document meant nothing. On the other hand, the constitution of UK is to a great extent not written; yet one may say that democratic procedures are strictly upheld by all involved.

That many people are required to establish a social fact immediately raises the question, how many? And further, if the number is not important, then what kinds of people are required? Suppose that two people marry in secrecy at a Swedish embassy. The only ones who know of the event are the two married people, the officiator and two witnesses. Nevertheless, it is a social fact that this couple is married, since the officiator was empowered, under certain conditions, to marry the couple. In this case, one can say that the social fact that these two people are married is established indirectly by collective agreement. We consider all people, regardless of whether or not we know them, to be married, if they acted within the conditions set by law.

There are interesting borderline cases where it is unclear which social fact is the case. A typical example concerns the power of a government during a revolution. Prior to the revolution, there are a number of social facts such as that the country has a government. During the revolutionary process, the established structure of power is dissolved, and after a while it is unclear what is the case. The remnants of the old government try to re-establish their position of power using propaganda and force, while the revolutionaries attempt to get rid of them, by force and other means. There is no obvious line to be drawn between the fact that the country has a government that is exercising its power and the fact that the old government and its power are no more.

The abundance of such borderline cases has prompted some people to abandon the concept of a social fact altogether. This is an unreasonable view. Many fruitful concepts are somewhat unclear and have many borderline cases. One such example is life. No one would doubt the fact that there are living things, and similarly that there are dead things. Nevertheless, it is difficult to draw a line in certain cases. For example, is a virus a living thing? Another borderline case is prions, i.e., proteins that are considered a likely cause of mad cow disease. Are prions living things, despite the fact that they lack DNA?

That there are many cases in which we cannot determine whether or not an alleged social fact is the case, does not mean that the distinction between brute facts and social facts is unclear. Rather, it is a consequence of the variable nature of social facts. Like all other changing things, many social facts come into being and vanish gradually, so the matter of their existence is sometimes unclear.

In discussing the difference between subjectivity and objectivity, one sometimes tends to forget that the line between them is drawn differently when dealing with natural phenomena as opposed to social phenomena. In the case of natural phenomena, a belief can be objectively false even though everyone shares it. It is objectively false if this belief does not agree with reality. But in the case of social facts, those facts are just collective beliefs and intentions. Therefore, the line between subjective and objective, with respect to social facts, is drawn between what one person believes and what many believe. A single person's belief about a social fact is subjective, but the common belief of many people is not. Rather, it is an objective fact about which a single person can be mistaken. This mistake is in regards to other people's attitudes, and is objective in the same sense as a mistake about a brute fact.

Knowledge about social facts is acquired by studying people's beliefs and the actions they perform in accordance with those beliefs. Naturally, this requires that we must interpret people's statements, actions and artefacts. There is simply nothing else to study. It could be thought that the result of beliefs and actions, e.g. written expressions, exist in a physical sense; and therefore, that we do not need to interpret anything, since it is sufficient merely to read the physically existing document. However, we cannot escape the problem of interpretation when a written expression is to be used in a concrete situation. Thus, knowledge about social facts is acquired through the interpretation of what people say, do or write.

5.9 Social Constructions

The term 'social fact' is not common in discussions of social scientific method and ontology. Instead, one often talks about *social constructions* (e.g. money, childhood, and in one book, reality[14]). What does this mean? As the term is normally used, it does not seem to have a single meaning. I shall give three interpretations.

(i) With the term 'social construction' one sometimes wants to emphasize that the phenomenon in question are created by a cooperative group of people. Given that one accepts that not everything is socially constructed, the term seems to have been introduced to mark a distinction similar to the one Searle makes between brute and social facts. However, some claim that everything is socially constructed, which yields a different interpretation of the term.

(ii) Sometimes people want to reject a statement as not truth-apt, with the argument that it is about a social construction. This argument presupposes that social constructions are not real. This can be interpreted as claiming that only physical objects and properties exist objectively. In my opinion, this view is incorrect. Propositions about constructed things, physical artefacts or social constructions, are just as true or false as propositions about non-constructed things.

(iii) Among those who use the concept, there is an implicit, and at times explicit, critique of the maxim that true sentences represent parts of reality. The distinction between reality and the linguistic representation thereof is refuted as untenable; thus making 'reality' a linguistic and therefore social construction. For example, one may hear the expression that people live in different realities, and that there is no observer-independent reality. An interesting confusion seems to be lingering here; for researchers and other people whose attention is directed towards natural phenomena normally use the word 'reality' to talk about the world regardless of how we perceive it. However, researchers in the humanities and social sciences often use a

[14] Berger, & Luckmann. *The social construction of reality*. New York: Anchor Books.

phenomenological interpretation of 'reality' along the lines of 'reality as it appears to someone'. Since our perceptions and theories regarding the world seem to be at least partially culturally dependent, it seems fair to say that reality (in this sense) is a social construction. However, Searle's point is, and I wholeheartedly agree, that even these constructions are real phenomena. They are embodied in humans (or in properties of humans) and can be articulated. Thus, one can talk about them, hypothesize about them, and make true or false propositions about them. In short, physical entities are not the only real entities.

Suppose that the constructivists persevere and argue that the claim presented above – that one could make objectively true or false statements about how people perceive the world – is also a social construction. This would mean that one cannot claim that a proposition is true in any other sense than that it is true for the person stating it; that is, 'true' means 'true for me'. Such a view is entirely relativistic and a modern parallel to the sophist Protagoras' view that everything is as it appears to the speaker. This view was previously discussed in Sect. 2.3.1, where it was rejected via Plato's counter-argument showing that relativism is self-contradictory.

If we assume that reality is a social construction, then we are want to ask, 'constructed from what?' According to Searle's analysis of social facts, we are to understand these constructions as attributions of certain functions/statuses to objects or actions. The objects to which we attribute a certain function can in turn also be social constructions. Thus arises the question: Of what is this object a construction? Apparently, we are left with an infinite regress, if we do not accept that there are things that are not constructed. Searle's discussion concerns social facts, but the same reasoning can be used in regards to social constructions. If the concept of social construction is to have any intelligible meaning, then there must exist things that are not social constructions.

In summary, I should say that – under certain interpretations – the concept of social construction seems to coincide with Searle's concept of social facts. In these interpretations, there are no problems with objectivity. However, if one attempts to develop a subjectivist or epistemological relativist view by claiming that everything is socially constructed, it is possible to refute this view in the same way that Plato refuted Protagoras' total relativism.

5.10 Criteria for Correct Interpretations

There does not seem to be a *general* criterion for correct interpretation. However, in some cases, like the example of Sweden's E.U. membership, it can be argued that *coherence* in the interpretation of all pertinent documents is the ultimate criterion. The interpretation of central clauses in these documents must lead to unanimity.

However, this does not exclude the possibility of conflicting coherent interpretations. This is a logical possibility, which in practice can be sidestepped given that one treats interpretation as an on-going process. Suppose someone were to propose a divergent interpretation of the meaning of Sweden's agreements with the E.U., and further claimed that this interpretation is coherent given that one reinterprets all other contracts regarding international cooperation etc. One could then argue that this interpretation is incorrect, since it breaks with the historical continuity of the interpretation of such contracts and other legal documents. Thus, the historical continuity of human and state relations becomes the distinguishing factor regarding different coherent, and yet conflicting interpretations. However, this does not solve the problem, since a revisionist interpretation of a text can certainly contain interpretations of earlier interpretations of that text. Thus, what is required is consistent revision. In the end, whether a certain interpretation of a text is correct will be decided by referendum, negotiation or by force. A good example illustrating this point is the Swedish judicial system, in which the Supreme Court has been mandated to interpret laws and the concepts applied therein. (It is the same in many other countries.) Thus some laws, or parts of laws, are construed by the interpretations made by the Supreme Court.

Another way of expressing this point is that the ultimate criterion for correct interpretation is determined by how all people involved comprehend their social relations. The coherence requirement therefore, in this case, translates to the requirement that the interpretation is accepted by (almost) all actors involved. This criterion is called the *actor's criterion*.

It is, however, not always the actor's, or actors', own comprehension that determines the ultimate criterion. Take, for example, one of Shakespeare's plays. When the play is to be performed at the theatre, the director must interpret the text in order to determine the point of the play. If we were to require the actor's criterion in such a case, then the author's intentions would be indicative of whether the director correctly interpreted the play. Of course, it is not necessary that director interpret the play exactly as Shakespeare intended. Thus, one cannot claim that the director misinterprets the play. However, this does not lead to any sort of relativism. The question 'what is the point of Shakespeare's play' has a definite answer (even if we can never know what it is), and the question 'what is the point of the director's version' may have a different definite answer. Naturally, the audience need not accept Shakespeare's nor the director's interpretation. For, the concrete performance of the play can be interpreted in a number of other ways.

The lack of a generally applicable criterion for correct interpretation is due, in part, to the large variety of uses one may have for a given interpretation, and partly to what Quine calls the *indeterminacy of translation*. It should be carefully noted that Quine's thesis about indeterminacy of translation is not the epistemological point that it is impossible to know the correct translation, but the more profound point that there are no facts of the matter determining correctness of translation. Quine has claimed, and many have agreed, that it is possible to produce multiple mutually exclusive translations of the same text, and nothing determines which is the correct one.

5.11 Summary

Qualitative methods are ultimately methods for making classifications of meaningful phenomena. When these classifications concern individuals, groups or institutions the resulting categories are interpretations of human actions and the artefacts produced by those actions. The question about the objectivity of these classifications is thus the question about the objectivity of interpretation.

Those who claim that all interpretation is subjective and that there are no objective facts in the human and social sciences have implicitly assumed that the distinction between objectivity and subjectivity is construed in the same way in all disciplines. This view is based on a confusion of two different meanings of objective and subjective. One can use the objective/subjective distinction to mark an ontological difference; certain phenomena are subjective in the ontological sense, if they are contained in, or part of, an individual's consciousness, otherwise they are objective. But we also use the objective/subjective distinction to distinguish between different types of propositions. A proposition is objective if it is true or false independent of what the speaker believes to be the case, otherwise it is subjective. Individual's mental states are ontologically subjective, but propositions *about* these subjective things are objective in the epistemological sense. A proposition is epistemologically objective if it is possible to establish criteria for the truth or falsity of that proposition and if different people's application of those criteria yield the same result.

Discussion Questions

1. A central concept in Economics is utility. The everyday word 'utility' seems to defy measurement, but Economics is predominately a quantitative discipline. Discuss what conditions are required to be able to quantify utility and whether these conditions are normally fulfilled.
2. A common view in Ethics is called utilitarianism, according to which we ought to choose the action that, in the long run, provides maximum happiness to as many people as possible. In order to realize such an ethical stance, one must be able to compare different states of happiness. Discuss the possibilities of both intrapersonal and interpersonal comparisons of states of happiness.

Further Reading

Berger, P., & Luckman, T. (1967). *The social construction of reality*. New York: Anchor Books.
Cresswell, J. W. (1998). *Qualitative inquiry and research design*. Thousand Oaks: Sage.
Glaser, B. G., & Strauss, A. L. (1967). *The discovery of grounded theory: Strategies for qualitative research*. New York: Aldine de Gruyter.
Glassner, B., & Moreno, J. D. (Eds.). (1989). *The qualitative-quantitative distinction in the social sciences*. Dordrecht: Kluwer.
Hacking, I. (1999). *Social construction of what?* Cambridge, MA: Harvard University Press.
Kirk, J., & Miller, M. L. (1986). *Reliability and validity in qualitative research*. Newbury Park: Sage.
Searle, J. (1996). *The construction of social reality*. London: Penguin.

Chapter 6
Theories About the Development of Science

'Anything goes.'
P. Feyerabend.

6.1 Introduction

The debate about the development of science is a debate about scientific rationality. The central question is whether there is any rationality in science, and if so, what that rationality could be. This question has long been discussed in one form or another.

Just after the First World War, this debate got a new focus in the work of the Vienna Circle. The starting point for this group was a general distrust of speculative metaphysics. Questions about reality's true nature, what lies beyond observable things – which were the traditional metaphysical questions – were considered meaningless.

Today, no one accepts the Vienna Circle's more specific doctrines; but even so, one may say that it had a large impact on the philosophical discussion about science. The analysis and critique of its fundamental points led to a fruitful development of concepts and methods within the philosophy of science. In this chapter I shall give an overview of four of the main positions on this topic: the Vienna Circle's program of logical positivism, Popper's falsificationism, Kuhn's theory of paradigms and scientific revolutions, and Lakatos' theory about research programs. I shall also give a short overview of Feyerabend's theoretical anarchism, and, finally, I shall propose a refinement of the concept of scientific rationality.

6.2 Logical Positivism

Beginning around 1920, a group of philosophers, sociologists, physicists and other academics met regularly in Vienna to discuss questions within science. This group, the Vienna Circle, focused their interest on the question of what is special about science. The natural sciences, especially physics, were taken as the ideal, with their

L.-G. Johansson, *Philosophy of Science for Scientists*, Springer Undergraduate Texts in Philosophy, DOI 10.1007/978-3-319-26551-3_6

great strides forward, while philosophy, and in particular metaphysics, trod much the same well-worn path it had been treading for over 2000 years. The Circle thought that one could establish a criterion that science usually fulfilled, but that speculative metaphysics did not: the *verifiability criterion*, which states that a sentence is meaningful if and only if one can describe a method to verify the truth of the sentence. (If it proved to be false, one would then have a method for showing that its negation is true.) It was claimed that many traditional metaphysical propositions, often the topic of philosophical discussion of the day, did not meet this criterion; and consequently, were strictly meaningless. In contrast, scientific propositions invariably did meet this criterion and were thus meaningful.

This criterion is a semantic criterion. It was an important part of a general theory of linguistic meaning to which the Vienna Circle subscribed. The main theses in this theory are that there are only two types of meaningful sentences; *analytical sentences*, which are true or false in virtue of the meaning of their words, and *synthetic sentences*, which are true if they are verified and false if their negations are verified. If a sentence cannot be placed in either of these categories, it is meaningless; they held this to be so even if it is syntactically well formed and where we believe we understand its meaning.

According to the Vienna Circle, how do we verify synthetic sentences? First, they thought that one could draw a sharp line between theoretical and observational sentences. Obviously, or so they thought, observational sentences are verified by making observations, the Vienna Circle saw no deep problem with this. However, the verification of theoretical sentences required a more worked out methodology.

The theoretical sentences in which the Vienna Circle had an interest were generally law-like sentences of the form 'all objects of type A have property B'. It was thought that such sentences are verified via an inductive method. An inductive method is a method that gives the conditions under which it is correct to conclude a general sentence (e.g. all swans are white) from singular sentences (e.g. swan 1 is white, swan 2 is white, etc.). This conclusion of a general law from individual cases was deemed legitimate if the following conditions are fulfilled:

- One has observed a large number of individual cases of the presumed law.
- One has not observed any counterexamples.
- One has observed these cases under many different circumstances.

It is obvious that the theoretical laws accepted as such in the natural sciences normally fulfil these conditions. However, these conditions are not very precise. There are two obvious problems: (i) exactly how many observations are required to say that a law is verified, and (ii) since one cannot, or should not, worry about all types of circumstance, which circumstances are the relevant ones? For example, suppose that on a winter day one observes that all the swans in Stockholm's waters are white. Can we now draw the conclusion that all swans are white? Before we do, it is reasonable to require that one investigate other places at different times of the year. On the other hand, it seems unreasonable to require that one take into account the weather or whether the king happened to be home at the Royal Castle as relevant circumstances. This is because we are convinced that these factors are not relevant to the colour that a bird's plumage takes. But how do we know? This knowledge is supposedly based on a general theory of causal connections in nature, according to which such factors cannot influence the colour of birds. However, this is theoretical knowledge that we have come by through other verifications, which, in turn, must also have met all three criteria above. Thus, we have gone round in a (Viennese) circle.

The Vienna Circle faced another difficulty in regards to some very general statements in science, such as the principle of the conservation of energy in closed systems. How does one verify such a statement? It is so general that it applies to absolutely every system, and so one begins to question whether it can be verified at all. But if such verification is impossible, then the principle of the conservation of energy must be a metaphysical sentence lacking meaning. This is not a reasonable conclusion. Thus, we are rather motivated to conclude that there is something wrong with verifiability as a criterion of meaningfulness.

The Vienna Circle, especially Rudolf Carnap,[1] reacted to this difficulty by changing the way they looked at theoretical sentences. Their view, developed in the 1930s, was that theoretical sentences are strictly meaningless, since they do not meet the verifiability criterion, and thus they have no truth-value. However, they are still used in science as logical tools for drawing conclusions about future observable events. The idea is that one first makes observations, formulates observational statements that are true and hence meaningful. Then, with help of correspondence rules and theoretical sentences (laws), one can logically infer new observational statements that describe situations not yet observed. Theoretical sentences in themselves do not claim anything. The theoretical terms we use do not refer to any unobserved objects. Theoretical sentences are mere instruments for prediction, which is why this view is called *instrumentalism*.

A necessary condition for holding this view is that one can maintain a very sharp distinction between theoretical and observational sentences, as they have completely different semantic properties. It is not enough to say that there is a vague line between theoretical and observational sentences, since this immediately raises the question as to whether sentences in the vague area have truth-values. This difficulty was one that the Vienna Circle could not satisfactorily solve.

Despite these glaring difficulties, this was 'the received view' up until the early 1960s. It is called logical positivism, or logical empiricism, and can be recognized as *verificationism in semantics, instrumentalism in regards to theories, and inductivism in regards to methodology.*

[1] Rudolf Carnap, (1891–1970), was born in Germany, but 1935 he moved to USA for political reasons. Carnap is the leading representative for logical positivism and heir to Frege, Russell and the early Wittgenstein.

6.3 Falsificationism

Karl Popper worked concurrently with many Vienna Circle members and sometimes participated in their meetings in the early 1920s. He agreed with some of their critical points in regards to metaphysical speculation and saw that it was important that one give a correct criterion for what can be called science. He was also critical of many popular theories of the time such as psychoanalysis and Marxism, whose supporters claimed were scientific theories. Popper argued that these supporters were never prepared to truly and unconditionally test their theories. However, Popper, as opposed to the members of the Vienna Circle, rejected any use of inductive reasoning in science. He argued that Hume had once and for all shown that the problem of induction was unsolvable. Hume's argument goes as follows. We often use inductive reasoning in science; we observe a number of events of type A each followed by an event of type B, then we generalise and conclude: all A-events are followed by B-events. In what way can this method be justified? It is obvious that this inference is not logically valid, so logical justification is out of the question. Then perhaps one might say the following: throughout the ages, we have found that inductive reasoning has quite often proved correct, even though one cannot claim any guarantee that the conclusions are true. Hence, this justifies further use of induction. But this argument is itself an example of induction; we have used induction in order to show that induction is a valid method, thereby assuming that which we set out to prove. Hence, we cannot justify induction using either our experience or logical arguments. But, then, what other method of justification is left? Nothing! For this reason, Hume argues, induction cannot be justified. Hume accepted that we humans naturally use inductive reasoning, in everyday life and in science, without justification, but Popper did not accept that. He held that science should be built upon rationally motivated modes of reasoning and hence we should reject all forms of induction.[2]

Popper's alternative was to build scientific reasoning on logically valid principles such as Modus Tollens. That is to say, if we infer an empirical consequence from a number of hypotheses and this consequence proves to be false, then we have a logically valid argument for the falsity of one or more of the hypotheses (See Chap. 3 and Appendix).

According to Popper, we are never justified in stating that we have proved some hypothesis, but we can say, under certain conditions, that we have conclusively falsified some hypothesis. This is the core of Popper's falsificationism. His methodology can be summarized in the following way:

- Put forth a hypothesis.
- Derive the consequences.
- Test the consequences and determine whether they are true of false.

---

[2] One should keep in mind that mathematical induction is quite another matter: it is an axiom for natural numbers.

- If any of the consequences are false, reject the hypothesis.
- If none of the consequences is false, then provisionally accept the hypothesis.

This is largely similar to the hypothetical-deductive method described in Chap. 3. However, there are some differences between Popper's view and the view I earlier expressed. In Chap. 3, I wrote, in agreement with most philosophers, that if the empirical consequences prove to be in agreement with experiment and observation, then one could say that the hypothesis has been strengthened. This strengthening can be interpreted as an increase of its believability, or equivalently, of the probability of it being true. This conclusion is inductive, since we argue from present knowledge to future unknown events. If the hypothesis is strengthened, then it is more probable than it was before the test that the next test will agree with the hypothesis. But Popper cannot claim that a hypothesis that has passed many tests has been strengthened, for then he would have to accept inductivism. Instead, he says that the hypothesis is *corroborated*. According to Popper, a theory is well corroborated if and only if it has survived many tests. It does not follow from the proposition that a hypothesis is well corroborated that the hypothesis is highly probable, or even that it is probable that the next attempt at falsifying it will fail.[3]

Popper also argues against all claims that a hypothesis is more probable after successfully passing a test with the following argument: Every general hypothesis (e.g. 'All swans are white') has a very large, and sometimes infinite, number of empirical consequences. The probability of a hypothesis is the quotient of a number of positive instances over the number of possible instances. If the number of possible cases is infinite, then it makes no difference how many observations and experiments with positive outcome one makes, because the quotient will always be zero. Thus, no finite number of tests can make a hypothesis more probable than if it had not been tested.

The question of how one 'discovers' good hypotheses cannot, and needs not, concern scientific methodology. Popper drew a sharp line between 'the context of discovery' and 'the context of justification'. According to him, there is no method for discovering or inventing new ideas since if there were, the results of such methods would not be genuinely new ideas or discoveries. What need justification are the conclusions drawn from such inspirations.

One problem with falsificationism is how it should treat the thesis that all observational statements are theory-dependent. This problem is related to the following: in order to say that one has falsified a hypothesis one must be able to say that the observation on which the falsification is based is irrefutable. However, if this observation, in turn, depends on the truth of certain theories, such as theories about our instruments or cognitive mechanisms involved, then one can ask whether it is the observational statement or the tested hypothesis that is false. But this means that we can never tell whether a hypothesis has been conclusively falsified.

[3] One should be well aware that Popper here uses the word 'corroborate' in his own technical sense. The usual meaning of 'corroborate' is 'support', or 'confirm'.

Popper's answer to this objection was to accept that observations are theory-dependent, whilst maintaining that before we even begin to test our hypotheses we should state what results we are prepared to accept as refutations. Thus, we are to point out a number of so-called base propositions that should not be questioned in a test situation (e.g. that certain instruments are reliable). Popper means that someone who is not prepared to make this methodological decision cannot be said to have a rational, scientific perspective. Obviously, this decision is not made once and for all; base statements, like all other scientific statements, should be open for critical discussion.

In my opinion this stance is entirely reasonable. However, for Popper, there is still the problem that he cannot consistently claim that it is possible to conclusively falsify a hypothesis. What he can say is that, *given* that we choose to treat a number of statements–those necessary for deriving consequences from a hypothesis–as true, we must *hold* that a hypothesis whose observational consequences conflict with observations is false. This is quite different from saying that the hypothesis in question has *been* falsified, since this would mean that we are certain that we know that it is false, and not just that we take it to be false, conditional on holding other propositions true.

Furthermore, there can be reason to point out that even the recognition of the need to make a methodological choice undermines Popper's strong requirements on rationality. For, on what grounds should we make the methodological choice of base propositions? Take, for example, the choice to consider a certain instrument to be reliable. Why should this choice be rational? Popper can hardly claim that it is based upon experiences of previous experiments, as this would be to use induction. Neither can a purely logical argument suffice, since logic alone cannot justify propositions about empirical facts. But then there is no justification at all. Hence the choice of base statements is, according to Popper's own norms, just as unjustifiable as every inductive inference.

In my view, Popper's falsificationism (he himself called his position 'critical rationalism') is not any real alternative to inductivism, since either he rejects all induction, thus making it impossible to motivate his methodology, or he accepts certain inductive inferences, which is essentially a variant of inductivism.

I think that induction is and will be an important part of all scientific thinking that aims at something more than merely describing events that have already has occurred. Popper's treatment of rational thought as being on par with strict logical thought and his simultaneous requirement for strict rationality in science results in science being reduced to pure logic, which, of course, is absurd. Nevertheless, Popper's methodology contains many reasonable points. As the reader surely has noticed, I have placed significant weight on the hypothetical-deductive method, and Popper should be given the honour of being one those who have worked hardest to spread this particular scientific method.

6.4 Normal Science, Scientific Revolutions and Paradigm Shifts

In 1962, Thomas Kuhn[4] published a book, *The Structure of Scientific Revolutions*, which quickly arose great interest among philosophers and other scholars. In his book, Kuhn presents his famous theory on the development of science as consisting of periods of normal science interrupted by scientific revolutions. The main concept in this theory is that of a *paradigm*, which has been widely dispersed, and has since come to be used in ways that Kuhn never intended. Thus, we must keep in mind that Kuhn's concept of a paradigm has a somewhat different meaning than is nowadays attributed to it, but more on this later.

Kuhn described the development of a certain science in the following way. In the beginning there was chaos; no established results to start with, no clear methods to apply. One did not even have a particularly well-described field of research or a list of problems to be solved. All researchers started from scratch with a fundamental philosophical/conceptual discussion and, since no results were indisputable, there was no foundation on which to build meaningful development. This phase is what Kuhn calls the pre-scientific stage.

Sooner or later a breakthrough occurs. Someone succeeds at solving a problem and everyone considers this success to be a clear step forward. Others then try to apply the same concepts, methods and techniques to solve new problems. If these attempts are generally successful, then one has taken oneself out of the pre-scientific stage and into a stage of normal science. No longer do researchers have to start with conceptual investigations, but can begin by arguing from the concepts, methods and results that others have already established. One is able to continue to build on this common ground taking ever more steps forward. This period of normal science is characterized by Kuhn as researchers working within a *paradigm*, where a paradigm is a complex of results, concepts, norms and assumptions. In his book, Kuhn used the word paradigm in many different ways. In fact, one critic counted 21 different definitions! In response to this critique, Kuhn, in the appendix to the second edition of his book, replaced his concept of a paradigm with that of disciplinary matrix, consisting of the following four components:

- Symbolic generalizations
- Metaphysical assumptions
- Methodological norms
- Exemplars

The last component, exemplars, is close to what Kuhn originally apparently intended by the word 'paradigm', and it is now generally accepted that the word 'paradigm' in a narrow sense refers to exemplars of successful problem-solving.

[4] Thomas Kuhn, (1922–1996) was an American philosopher and historian of science.

An example may illustrate the meaning of these terms. Newton's *Principia* was immediately considered a breakthrough in mechanics, the science of bodily motion. It quickly became a paradigm for both mechanics and physics in general, and remained so for more than 200 years.

The symbolic generalizations in this normal science are Newton's laws. Among the fundamental metaphysical assumptions, we find Newton's assumptions about absolute space and absolute time. Another quite important metaphysical assumption is determinism: the assumption that Newton's laws, together with a set of initial conditions, determine all bodily motion.

Among the methodological norms, we can name the norm that hypotheses should be strengthened through systematic experimentation (though Newton probably had a somewhat different view in this case) and that physics should use mathematics in order to make numerically precise propositions about the motion of bodies.

There are many prime exemplars in *Principia*, such as Newton's analysis of lunar motion as a continual free-fall caused by the earth's gravitation and his analysis of the motion of the tides as an effect of gravitational forces from the moon and the sun.

Once a paradigm has been established in a discipline, it functions as a general conceptual framework. Researchers are more or less unconscious of the framework and its governing power. The paradigm contains the concepts and assumptions that researchers take to be so obvious that they never reflect upon them or consider their justification. The common reaction to a researcher who questions any part of the paradigm, by their colleagues, would be to question his or her ability or judgment rather than questioning the paradigm. This reaction is even more pronounced where the critic is not a researcher within the paradigm in question. Consequently, and this is Kuhn's point, as long as normal scientific activity is reasonably productive, the paradigm is never questioned.

Working within a normal science may appear to be a trivial and unimaginative activity, but such is not the case. Creativity and fantasy play a significant role, but that role is defined by the paradigm.

Sooner or later the researchers of a normal science run into some difficulties. Kuhn claims, quite reasonably, that unsolved problems always exist in all sciences, but as long as these problems are not too many, or too insistent, then they are not viewed as grounds for doubting the fundamental assumptions of the paradigm. However, if the number of unsolved problems–which Kuhn calls anomalies–become sufficiently serious and numerous, researchers in the field will begin to question the fundamental assumptions. One becomes aware of certain aspects of the paradigm in question, and in light of the accumulation of anomalies, one begins to question these aspects. The result is that the discipline enters a crisis. This, in turn, is a precondition for a scientific revolution and for throwing out one paradigm for another. The new paradigm may be the conceptual framework for a new period of normal science.

Kuhn's most controversial thesis is that different paradigms within a discipline are *incommensurable*, or incomparable. Kuhn has two arguments for this thesis.

The first is that different paradigms use different concepts in describing a research field. The meaning of these concepts is decided holistically: each concept's meaning depends upon its relation to other concepts in the paradigm. This view is called *semantic holism*. Semantic holism applies even to the concepts that we use to describe our observations, which means that all observations are theory-dependent. This, in turn, means that two propositions claimed within two different paradigms cannot in general be said to be about the same thing; and therefore, can never be said to be in conflict.

The second argument builds upon the fact that norms are part of paradigms. If one is to objectively compare two paradigms, then one must have some criteria or scale against which we measure concurrent paradigms' relative strengths and weaknesses. However, using such criteria is to apply methodological norms and such norms comprise significant parts of paradigms. If the methodological norms of the two paradigms are different, we cannot give an impartial verdict, for it is obvious that we have (partly) already chosen a paradigm when we choose which methodological norms are acceptable in comparing the two paradigms. One can metaphorically describe this situation as there not existing a neutral ground on which one can stand and impartially compare to concurrent paradigms' strengths and weaknesses.

However, if one cannot compare two paradigms, one established and the other new, then how do we choose between them? Apparently, one cannot make a rational decision based on a comparison of advantages and disadvantages from a neutral point of view. Yet, the scientific community, the active researchers within a discipline, do make a choice. According to Kuhn, this decision must be based on external factors such as 'political correctness', personal sympathies and antipathies, the possibility of getting funding, etc. The main point here is, given one's goal is to make scientific progress, there is no rational way to make a choice such that one optimizes the chances of attaining that goal, if Kuhn's incommensurability thesis is correct.

Kuhn's critics think this is absurd: if science is not the pattern of rational decision-making, then nothing is. Thus, there must be a rational methodology in science and Kuhn must be wrong.

Kuhn did not fully accept this consequence that paradigm choice is irrational. He maintained that his theory did not lead to irrationality. In my opinion however, maintaining that science is fully rational is nigh on impossible, if one accepts, as Kuhn does, that two different paradigms are incommensurable.

Fortunately, one can question the strength of the premises on which the incommensurability thesis is built. In my opinion, Kuhn has greatly exaggerated the extent and implications of semantic holism. For instance, Quine–who accepts a strong form of semantic holism–has not drawn the conclusion that it is impossible to compare and choose between theories. As discussed in Chap. 5, Quine thinks that there are theory-independent statements that, in a certain sense, constitute the foundation of science. Thus, he maintains that one can objectively compare theories at this level. Despite this difference, Quine agrees with Kuhn that holism implies that individual statements, or sentences, in a theory cannot be tested in isolation, but that one must test and compare entire theories.

Several philosophers, less holistically inclined than Kuhn and Quine, have used Frege's distinction between a term's meaning and reference. Using that distinction one may accept that a term has changed meaning, while maintaining that its reference remains, and this enables comparisons between paradigms.

Even the second argument is strongly exaggerated. Many of the paradigm shifts that Kuhn discusses in his book are not such that the scientific norms have changed to any significant extent. In most cases, when a scientific revolution and paradigm shift occur, the major norms remain unchanged in regards to, for example, the value of accurate predictions. This is why one often can compare new and old paradigms.

In summary, Kuhn's arguments for the most controversial thesis in his theory, the incommensurability of paradigms, are weak and provide insufficient support. Consequently, there is no basis for claiming that the conclusions drawn during scientific revolutions are as irrational as Kuhn's theory suggests.

Kuhn has in later publications[5] developed and modified his views about scientific change, incommensurability, paradigms and rationality, but these writings have been less influential.

6.5 Lakatos' Theory of Research Programmes

Imre Lakatos[6] succeeded Popper as professor of philosophy of science at the London School of Economics and was a follower of Popper in regards to his stance concerning questions about scientific development.

Lakatos called his theory 'methodology of scientific research programmes' MSRP for short. One can see this theory as an attempt at saving as much as possible of Popper's view of science whilst at the same time taking into consideration the objections raised against it. In some respects, Lakatos was also influenced by Kuhn's ideas on scientific revolutions, though he did not accept Kuhn's incommensurability thesis. Lakatos – like Kuhn, but unlike Popper – was strongly oriented towards the actual, historical development of science in developing his methodology.

Lakatos had two starting points from which he constructed his theory of research programmes. The first is the observation that certain large scientific breakthroughs are actually best characterized as verifications and not falsifications, as Popper had claimed. Characteristically, these verifications are such that two theories are compared using a decisive experiment. Lakatos illustrated his thesis with Eddington's observations of the deflection of light at the sun's edge, which was carried out during the solar eclipse of 1919. During a solar eclipse, one can see the light from

[5] Collected in his Kuhn (2000).

[6] Imre Lakatos (1922–1974) was born in Hungary, but fled to England in 1956.

stars that pass near the edge of the sun, on its way to earth. Einstein's general theory of relativity predicted that the sun would bend this light as it passed by the sun. The results of Eddington's observations agreed with relativity theory and conflicted with classical mechanics. According to Lakatos' interpretation, it was clear that this episode verified relativity theory while at the same time falsifying classical mechanics.

Lakatos' second starting point is that one neither can, nor should, evaluate a single theory in regards to a fixed point in time; rather, all that one can evaluate are tendencies in the development of a series of theories (i.e. a research programme). The deciding question is whether the research programme is progressive or degenerative. In order to determine which is the case one must adopt a historical perspective.

According to Lakatos, a research programme consists of a series of related theories. The common core of such a series is called the *hard core*. This concept is quite similar to Kuhn's concept of symbolic generalizations. Furthermore, there is a *protective belt*, which consists of assumptions about measuring instruments, observation conditions, background assumptions, etc. The purpose of the protective belt is to provide a buffer between the hard-core and recalcitrant experimental results; instead of saying that a result shows that the hard core is false, one says that some assumption in the protective belt is false. This principle is a heuristic principle, the negative heuristic, which Lakatos formulated as a categorical prohibition; do not use a Modus Tollens argument against the hard core! Finally, there is also a *positive heuristic*, which consists of guidelines for the construction of new theories in the research programme. Thus, a research programme has four components:

- The hard core
- The protective belt
- The negative heuristic
- The positive heuristic

Lakatos' basic methodological rule is the following: give up a degenerative research programme in favour of a progressive research programme. A degenerative research programme, according to Lakatos, is characterized by theory development that occurs primarily in response to empirical difficulties; one finds new alternatives, exceptions, or reinterprets results in order to explain why the original theory no longer agrees with the observations. A progressive research programme, on the other hand, is characterized by theory development that pre-empts experimental results. Theories in the research programme have inherent possibilities for further development, which able researchers take advantage of prior to experiments taking place to test these possibilities.

One cannot reasonably say that one should give up a research programme at the first sign of difficulty. That would be too rash. Therefore, Lakatos argued that the methodological rule of throwing out a degenerative research programme should be supplemented with the regulation that one should give a research programme that has come into a degenerative phase a certain amount of time to regain its force and once again become progressive. This immediately triggers the question, how long time? Lakatos gave no answer to this question, which leads one to concur with the following critique, as formulated by Feyerabend: Lakatos has not, in all practicality, given us any methodological rule to follow. Using Lakatos' rule, every conclusion can be defended as rational, which means that he has not succeeded in distinguishing between rational and irrational methods. Without breaking Lakatos' rules, one can both claim that it is rational to give up a research programme because it is degenerative and that one should stick with it because it is in a phase of temporary difficulty. Furthermore, there are well known historical examples that show that it can take several decades before one is able to overcome temporary difficulties. As a result, Feyerabend characterized Lakatos' methodology as 'verbal ornamentation'.

A response to this critique can be reconstructed from Lakatos' remarks and general positions (since Lakatos died in the midst of this debate he did not himself respond) in the following manner: The theory about research programs is not primarily intended as a methodological rulebook for on-going research. Instead, it is to be seen as a methodology for *rational reconstruction* of the history of science. When we look back at what has happened in the history of science, it is possible, according to Lakatos, to describe the development of events as a rational process, since the conclusions made in practice fit his model. When programmes were abandoned, it was because they had been in a degenerative phase for some time, and when programmes were not abandoned, it was because one was trying to overcome temporary difficulties. *With hindsight,* we can say that large parts of the history or science are rational processes. However, one cannot expect that individual researchers would be more rational than other people would. Researchers – like most people – indulge in wishful thinking, get caught up in impossible views, and let short-sighted career interests influence their conclusions, etc. However, this does not take away from the rational character of their *collective activity*.

Lakatos is, in contrast to most in the debate about scientific development, a methodological collectivist (see Sect. 9.1) in that he thinks one should explain each individual's decisions and beliefs in terms of objectively acting forces. The general idea of methodological collectivism was first proposed by Marx and in his theory these objectively acting forces are social in nature, (in Marx' theory they are the conditions for production) but when talking about scientific research one can interpret them as nature exerting a selective pressure on our perceptions, decisions and ideas. Individual researchers are not more rational than other people, but the research community is influenced, as a whole, by objective reality; and hence, the scientific community now and then changes research programme. Lakatos' concept of rationality should thus be used primarily at the collective level. The critique of Lakatos that he has not given any solid rules that an individual researcher can follow in choosing between research programmes misses the point, for this was never Lakatos' goal.

Let me stress once more that Lakatos has not explicitly expressed these points himself; rather, this is an interpretation of his position. It is Ian Hacking in particular (in his book *Representing and Intervening*) who has pushed this interpretation. I find it plausible since Lakatos, as a former communist, was strongly influenced by Hegel and Marx, and both of these thinkers were methodological collectivists.

6.6 Methodological Anarchism: Anything Goes

Paul Feyerabend[7] participated intensely in the debate about the scientific method and adhered to essential parts of Kuhn's view. In particular, he wholly accepted the incommensurability thesis. Though he went one step further than Kuhn in criticizing the popular philosophical position that what we know to be science employs a single identifiable method and the success of that method explains scientific progress. He has partly a descriptive and partly a normative critique of those, like Popper, who maintain that there is only one general methodology that qualifies as scientific. His first point concerns the history of science. Each proposal for a scientific method hitherto considered has, in practice, often been neglected or actively rejected by active researchers, and yet these researchers have managed to make progress. Feyerabend claims that if researchers were to accept any one of the philosophers' proposed methods, it would hinder scientific progress.

Feyerabend's way of looking at science is quite similar to the ultraliberal economist's view of capitalism and market economy. These economists claim that the most effective way to promote a rapid increase in wealth is to abolish all, or most, regulations and let competition in the market take care of itself. The products and services that best satisfy the customer's desires will be produced, and every attempt to direct the market will diminish its effectiveness.

Feyerabend argues that the same goes for research goals and methods. He even argues that the state should stop favouring activities traditionally called 'research'. One should allow all possible traditions of acquiring 'knowledge' such as Voodoo, astrology, crystal therapy, homeopathy and the like, to use a part of societies' resources for equal competition. This will speed up scientific development and increase the chances of dispersion of radically new insights. Finally, Feyerabend argues that there is no reason to accept the traditional view that knowledge, understood in the traditional way, leads to a happier life for human beings. In a truly liberal society, everyone has the right to decide his or her life's goal. So, what would the majority prioritize if they could choose? Perhaps they would not want to promote science? In short, the only methodological rule that Feyerabend accepts is that there is no binding methodological rule; *anything goes*.

Some of the Feyerabend's wording bears traces of the rhetorical figures and ideological debate that was prevalent, especially at universities, during the 1960s and 1970s. For my part, I find his analysis of the history of science quite relevant

[7] Paul Feyerabend (1924–1998) was born in Austria, but was mostly active in USA.

and pertinent. Without having investigated the issue more systematically, it strikes me as an accurate observation that science has made progress without consciously following any methodological rules. It is also clear that no hitherto proposed rules have been simultaneously generally accepted and practiced; and yet progress has been, and continues to be made. Furthermore, I think that philosophers should be extremely careful when prescribing how one should act in order to be a rational, genuine scientist.

However, if one accepts the view that progress occurs if we let theories and research projects compete in the 'market of ideas', then one needs an explanation of what is meant by 'progress'; why one effort result is considered progress, but not another. Feyerabend, like Kuhn, is faced with difficulties regarding the concept of progress due to his belief in the incommensurability thesis. The reason is that in order to claim that progress has occurred it does not suffice to say that a new theory is different from an older one; that the newer theory represent progress presumes that one has compared the two theories with respect to some criteria that the new theory better satisfies. If a comparison has been made, it means that the theories are not incommensurable. I would even go so far to claim that it is self-contradictory to simultaneously hold that progress has occurred when one theory is given over for another and that those theories (paradigms, or research programmes) are incommensurable.

Feyerabend's stance that scientific knowledge, traditionally understood, does not lead to happier life for people in general, I strongly dismiss; it strikes me as absurd to deny that common people in the developed countries live a better life nowadays than, say 200 years ago, and that this is due mainly to scientific achievements.

6.7 Summary of the Debate

This chapter gives a short overview of twentieth century currents in the study of scientific methodology. The main issue is, and has always been, how we should characterize scientific methods in such a way that one can use the characterization to exclude from science all those activities that do not properly belong to science (superstitions, religious arguments, pseudoscience, etc.), and include in science all those activities that do so belong (physics, biology, astronomy, etc.). This formulation suggests that there is a tension between the normative and descriptive goals of philosophy of science. According to my grasp of the situation, Popper's biggest mistake was that he was far too normative and did not sufficiently take into consideration how science is actually performed. This mistake allowed his critics to question the relevance of his view. Kuhn, on the other hand, made the opposite mistake. He concentrated far too much on describing the development of science, which lost him the possibility of saying anything about how science ought to be carried out in order for it to qualify as science. Philosophical theories about the development of science must balance both of these aspects in order for them to have any relevance.

Accordingly, I propose that the starting point should be the hypothetical-deductive method and the related approach to scientific theories. A generally held characteristic of what 'scientific' means could be the following:

> To put forth hypotheses, infer testable consequences of those hypotheses, and to be prepared to reassess one's hypotheses if the consequences do not agree with observation.

This formulation is quite similar to Popper's falsificationism. The difference is that Popper more precisely claimed that a scientist *should* abandon a hypothesis if the consequences conflict with observation, whereas I think that such a formulation is far too categorical. There have been many cases in the history of science where one has stood by a hypothesis even though it seemed to conflict with observation. This strategy has shown itself to be correct on occasion and has generally been considered both scientific and sensible. It is also easy to cite basic arguments for such a stance. One argument is that verified observational statements are not unshakable truths. On any occasion, there could have been something wrong with the measuring instruments, or there might have been some unknown interference in connection with the observation. Another way of expressing the same thing is to point out the possibility that some auxiliary assumptions may have been false. Therefore, I think that one cannot categorically state that one should abandon a hypothesis as soon as it comes in conflict with observation, but rather one must make a comprehensive comparison of the probability that something is wrong with the hypothesis and the probability that something is wrong with the observations, and then conclude which seems the most reasonable. (Popper later modified his view, holding that observational statements, which he called base statements, can and should be subjected to critical inspection, just like the rest of a theory.) Adherents of pseudoscience would now be able to claim the following:

> There is no clear difference between our methods and your so-called scientific methods. We also put forth hypotheses that lead to observable consequences, and you are just as dogmatic when you stand by your hypotheses in the face of recalcitrant observations as we are. In practice, the only differences are the principles we believe in. Your attempts to demarcate what is scientific (namely, your own methods) are merely a privileged group's attempt to hold on to its privileges, its grants and society's appreciation. In a liberal democracy one should allow all people to do what they want, as long as it does not hurt others. Since proponents of so-called scientific methods have not succeeded in showing that these methods lead, in the long run, to more secure progress and better results, it is entirely possible that we would be able to make significant progress if only it were allowed that more ways of looking at reality were given the resources to investigate their inherent possibilities. (This is essentially Feyerabend's argument.)

This critique of established science cannot be immediately rejected. Unfortunately, the first argument–that scientists have not succeeded in presenting a sufficiently clear criterion for what science is–is true. My approach above is not precise enough to withstand the critique. The second argument also has a certain air of truth about it. If one is in academia, one is dependent on access to research grants, and it is natural that one would try to reduce the competition by claiming that potential competitors do not fulfil the requirements of science.

Nevertheless, I am convinced that there is a sizable difference between science and pseudoscience. There are certainly difficult borderline cases, but there are also clear-cut examples on both sides of this line. This difference has to do with the way a research collective evaluates and revises its beliefs. It can very well be that an individual researcher breaks with scientific norms, but this is not reason enough to reject an entire activity as unscientific. The relevant level at which to decide this issue is not at the individual level, but at the collective level. Three reasons for this are (i) what counts as scientific results is not what an individual researcher or research group claim but what those who are familiar with the subject, the researchers at the forefront of their field, accept as a whole, (ii) at the collective level, the individual researchers' private opinions regarding politics, ethics, religion, and their career interests will cancel each other out when it comes to the collective decision, and (iii) when faced with a conflict between hypothesis and observation, it is not the individual researcher's judgment that determines which conclusion will be drawn, but that of the entire research collective. For example, in the case of a conflict between theory and experiment, other researchers repeat the experiment under different circumstances in order to try to determine if the observations are at fault, or if the observations are correct and it is the hypothesis that requires revision.

Hence, I would argue that one could harbour certain expectations regarding the rationality of science, even though individual researchers are just as irrational as everyone else. The crucial question is what the institutional organization looks like. It is required, among other things, that there are multiple independent research groups working within the same subject area, that these groups have the possibility of sharing results and that these results are published and subjected to debate and critique by others within the subject area. In other words, there are reasons to believe that it is possible to define a concept of scientific rationality, even if it is impossible to give a sharp delineation between science and pseudoscience.

6.8 The Rationality of Science: A Model

William Newton-Smith has proposed a model of scientific rationality in his book *Rationality of Science*. It is intended as an analysis of theoretically interesting cases where researchers have abandoned an old theory for a new one. This change of beliefs is, according to Newton-Smith, a rational process if it fulfils the following criteria:

1. The scientific community has a goal specified by the model with its scientific activity.
2. Between two competing theories T_1 and T_2, one of them, say T_2, is closer to this goal, given the available evidence.
3. The scientific community realizes that T_2 is closer to the goal than T_1.

4. The scientific community rejects T_1 in favour of T_2 on the basis of the realization that T_2 is closer to the goal than T_1.

This model is nothing but a variation of the model for action explanations given in Chap. 8, this time applied to collective decisions and collective beliefs; and as I have previously stated, there is every reason to discuss scientific rationality at this level.

I have, however, a minor objection to this model. One cannot say that all activities that fit into this model are scientific, because the goals are not specified. For example, if one were to plug in for the goal an increase in profitable activity, this model would apply to sales-oriented companies. (Theories would then become systems of assumptions about business plans, markets, etc., and the scientific community would become upper management.) What is specific to science is, I would say, the search for truth. (This is not explicit in Newton-Smith's view.) Many would try to fend off such a solemn formulation, preferring formulations such as to explain what happens, to predict what will happen, to find a theory that is empirically adequate (van Fraassen), or to find useful theories. In my opinion, all of these formulations strike me as unsatisfactory on the basis that they do not sufficiently distinguish science from other activities that should not reasonably be called science.

Point 3 above assumes that T_1 and T_2 are comparable with respect to how close they are to the goal. This is a notoriously difficult problem in the philosophy of science where one takes the goal to be truth. Popper tried to define closeness to the truth, *verisimilitude*, in terms of sets of true and false sentences in the theories in question. This attempt was a capital failure, and many of Popper's critics have taken this to indicate that Popper's philosophy is bankrupt. Regardless of whether or not this conclusion is justified, I believe that no one who wants to discuss scientific methods and progress can do so without an ability to judge the relative merits of competing theories. Of course, one can say, formally, that a theory, which contains a single false sentence, is false and should be abandoned. Consequently, as it is likely that all theories we can come up will be marred by some falsities, so all scientific theories will prove equally false and so equally worthy of rejection. This is not a reasonable stance, there are better and worse theories and a bad theory is better than no theory at all. Ilkka Niinilouto has discussed this issue in depth[8] and has defined a concept of truth-likeness in terms of quantitative differences between theoretical predictions and observations. This seems to be an intuitively reasonable way to go.

Finally, point 4 is a clear expression of an internalistic position. If it were the case that the relevant scientific community abandoned one theory in favour of another on the basis of other reasons than that it is closer to the goal (e.g. political, economic or religious reasons), then we would not say it was a rational step, even if the decision could afterwards be motivated by a high

[8] Ilkka Niiniluoto (1987). *Truthlikeness*. Dordrecht: Reidel.

probability of fulfilling the goal (e.g. acquiring large research grants for active researchers). This is entirely analogous to how we view an individual person's actions. If someone who reads in a horoscope that he/she could have success in financial dealings becomes inspired to buy a lotto ticket and wins millions, we would still hardly say that this action was rational even if the action, given the agent's individual goals, led to a high degree of goal fulfilment.

Exercise

Below is a fictional dialogue between Doctor Faustus, the medieval scholar who, according to a classic German legend, sold his soul to the devil for knowledge, and a present-day physics student. Perform a science-theoretical analysis of Faustus' argumentation!

Friction Dialogue

Participants:
Doctor Faust, a medieval scholar in league with the devil.
Pelle, a present-day physics student.

F You are always talking about friction. How do you know that it is friction that stops a ball from rolling and not demons?
P I do not believe in demons.
F I do.
P Anyway, I do not see how demons can make friction.
F They just stand in front of things and push to stop them from moving.
P I cannot see any demons even on the roughest of table.
F They are too small, also transparent.
P But there is more friction on rough surfaces.
F More demons.
P Oil helps.
F Oil drowns demons.
P But if you polish a table there is less friction and balls roll farther.
F You are wiping the demons off; there are fewer to push.
P A heavier ball experiences more friction.
F More demons push it; and it crushes their bones more.
P If I put a rough brick on the table I can push against friction with more and more force up to a limit, and the block stays still, with friction just balancing my push.
F Of course, the demons push just hard enough to stop you moving the brick; but there is a limit to their strength beyond which they collapse.
P But when I push hard enough and get the brick moving, there is a friction that drags the brick as it moves along.
F Yes, once they have collapsed the demons are crushed by the brick. It is their crackling bones that oppose the sliding.
P I cannot feel them.
F Run your finger along the table.
P

Friction follows definite laws. For example, experiment shows that a brick sliding along the table is dragged by friction with a force independent of velocity.

F Of course, same number of demons to crush, however fast you run over them.

P If I slide a brick along the table again and again, the same friction is the same each time. Demons would be crushed in the first trip.

F Yes, but they multiply incredibly fast.

P There are other laws of frictions: for example, the drag is proportional to the pressure holding the surfaces together.

F The Demons live in the pores of surface: more pressure makes more of them rush out to push and be crushed. Demons act in just the right way to push and drag with the forces you find in your experiments.

Adapted from Eric M Rogers: *Physics for the Inquiring Mind.*

Further Reading

Feyerabend, P. (1988). *Against method.* London: Verso.

Hacking, I. (1983). *Representing and intervening*. Cambridge: Cambridge University Press.

Kuhn, T. (1962). *The structure of scientific revolutions*. Chicago: Chicago University Press.

Kuhn, T. S. (2000). *The road since structure: philosophical essays, 1970–1993*. Chicago: University of Chicago Press.

Lakatos, I., & Musgrave, A. (Eds.). (1970). *Criticism and the growth of knowledge*. Cambridge: Cambridge University Press.

Newton-Smith, W. (1981). *Rationality of science*. London: Routledge.

Popper, K. (1992[1959]). *The logic of scientific discovery*. London: Routledge.

Part II
Philosophical Reflections on Four Core Concepts in Science: Causes, Explanations, Laws and Models

Chapter 7
On Causes and Correlations

> *Development of Western science is based on two great achievements: the invention of the formal logical system (in Euclidian geometry) by the Greek philosophers, and the discovery of the possibility to find causal relationships by systematic experiment (during the Renaissance).*
> *A. Einstein*

7.1 Causes Are INUS Conditions

When seeking a scientific explanation of a given phenomenon, we often look for its cause. We ask for the reason for why we have caught a cold and why the boiling-plate becomes hot after having switched on the electricity.

In situations such as these, if we possess a certain level of general knowledge of scientific matters we can conclude that a cold is typically caused by a virus and that the boiling plate becomes hot due to the friction of electrons in the conductive lines which reside within the metal. But what does it mean for one event to be the cause of another? A preliminary answer would be that the cause of an event is that which makes the event take place. But what do we mean by 'makes'? Is it not more or less synonymous with the word 'cause'? Have we really improved our understanding of causation just by replacing 'makes' for 'causes'? Furthermore, does each event have a single cause, or are there circumstances in which an event may have more than one cause?

People often say that a cause *necessitates* a certain effect. What do we actually mean by this? It cannot really be a matter of logical necessity, as the relationship between cause and effect is not purely logical. David Hume (1711–1776) argued against the idea of a necessary relationship between cause and effect, if this notion is taken to mean some sort of necessity in the nature. He asked what it is we in fact observe when we observe that a cause is followed by its effect. According to Hume, all we can observe is the following: (i) cause precedes effect, (ii) cause and effect are contiguous, and (iii) there is a constant correlation so that each time we observe the same type of cause, we can also find the same type of effect. Thus there is no logical connection between cause and effect, nor can we observe any other kind of necessary connection. Hume therefore concluded that our idea of a necessary connection between cause and effect is a strictly psychological habit, arising after

L.-G. Johansson, *Philosophy of Science for Scientists*, Springer Undergraduate Texts in Philosophy, DOI 10.1007/978-3-319-26551-3_7

having regularly seen a certain type of event followed by another type of event. We become conditioned to expect that a subsequent event of a certain type will follow from a prior event of another type, and it is this expectation that creates our belief in a necessary correlation between cause and effect. Strictly speaking, no cause is, in a modal or metaphysical sense, necessary. It is just a matter or regularity.

Many people have critically opposed this negative conclusion, yet many others have also accepted Hume's view as essentially correct. John Mackie is a modern philosopher who has further developed the notion of causality in the spirit of David Hume. In his book *The Cement of Universe*, Mackie defines a cause as a so-called *inus-condition*. In order to explain what this means, Mackie use a house fire as example. After such an occurrence, most people, especially the insurance company, would want to find out what caused the fire. The local Fire Department normally conducts a formal investigation, to determine the cause of the fire, so let us assume that the investigation concludes that the fire was caused by a short circuit in some faulty electrical wiring. A spark created at the time of short lit up nearby curtains and since nobody was home to put out the fire or call the Fire Department, the whole house was destroyed.

Based on this conclusion we can list several necessary factors that contributed to this event; the wires were in poor condition, the main circuit breaker for the house was 'on', the curtains were too close to the point in question, nobody was home, there was oxygen in the room, and so forth. It is evident that if one of these circumstances had not been present, the house would never have burnt down. Why did the fire engineer consider the bad wires as the real cause of the event? Why not the curtains? Why did he not claim that the cause of the fire was that the main circuit breaker was on, or that there was oxygen in the room? Each of these conditions is equally necessary for event's occurrence.

Intuitively, it is clear that in this case we would never say that the presence of oxygen, or the main power switch, or any of the other circumstances listed above caused the fire. The reason for this is probably that circumstances that appear natural and normal do not cause deviations from the normal course of events. It seems as if our thoughts follow the principle 'abnormal effects follow abnormal causes'. However, what is considered abnormal is certainly a function of our personal perspective and our personal experiences. What may be considered normal or abnormal in specific situation is not merely an objective question. The conclusion of this argument is that a 'cause' is one of the necessary conditions for a particular occurrence, but it is impossible to objectively determine which of the necessary conditions actually caused the event.

Furthermore, the concept 'necessary conditions' (see Appendix; *necessary condition* is *not* a modal concept!) implies that in order to generate a particular effect, all necessary conditions must obtain. We can therefore conclude that a cause is a necessary component of a complex set of conditions, which together bring about a certain effect.

It is, however, obvious that a house can burn down due to other circumstances other than a short circuit in the electrical system. Fires can be caused by pyromaniacs, lightning, or sparks from nearby fires. Thus the set of conditions, which together caused the house to burn, is one of many possible sets. With this point in mind, one may offer the following definition of the concept of a cause:

> *Def.* A cause is a necessary component of a set of conditions, which together are sufficient for the occurrence of some effect.

This is a somewhat simplified version of John Mackie's definition of a cause, expressed in terms of the so-called *inus conditions*:

> A cause is an Insufficient but Necessary part of an Unnecessary but Sufficient condition for the effect.

I explained above that a selection criterion for a given cause is that it must deviate from what would be considered normal. Another basis for selecting a particular inus-condition as the cause of a given event is the possibility of manipulation. A third reason to refer to an inus-condition as a cause is that it has a high degree of specificity.

There are many medical examples that illustrate this point, such as the cause of the common cold. It is well known that the cause of the flu, or the common cold, is a specific type of virus. However, all who are exposed to the virus do not develop the illness. For the virus to cause an infection one needs to have a weakened immune system. So why can we not just say that the flu is caused by a weakened immune system? Well, that would be reasonable, but we typically point to a virus being the cause of a common cold or the flu because the virus has a specific effect. If the immune system is compromised, several illnesses can take hold, or, if you are lucky, you may not contract any.

We can easily imagine a somewhat different biological system where the flu virus is always in the environment and that we normally do not become ill, except for when the immune system has been compromised. In that case we would probably say that the cause of the flu is a weakened immune system.

In the medical field one often make a distinction between genetic causes and environmental effects. As the following example[1] will show, this distinction is not that clear.

Some breed of the species *Drosophila Melanogaster*, the common fruit fly, has a genetic defect that results in its wings being significantly shorter than normal. This occurs if the temperature is around 20° Celsius at the time the fly is maturing. But if we change the temperature to about 32 °C, we find that the wings grow to normal length. What is the cause of the shortened wings? (Fig. 7.1)

The answer to this depends on which contrast is being considered: if we compare two populations with the same defective gene pool at different temperatures we would naturally conclude that it is the low temperature which causes the shortened wings. However, if we compare two populations of flies at the same temperature, of which one has shortened wings and another does not, we would say that it is the genetic factor that caused the shorter wings.

---

[1] The example is taken from Hesslow's, G. (1984). In Lindahl & Nordenfelt (Eds.), *What is a genetic disease? On the relative importance of causes* (pp. 183–93).

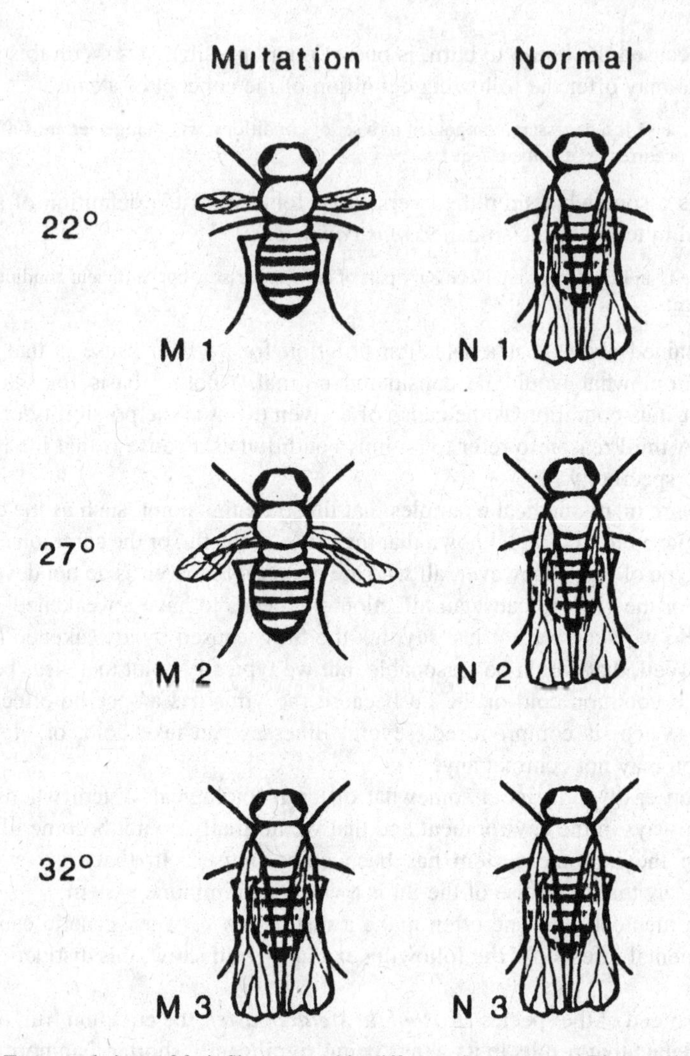

Fig. 7.1 Two populations of fruit flies matured in different circumstances (Picture adapted from Hesslow 1984)

Far from all defects which we typically consider to be genetic can be compensated for by manipulating environmental factors, but the example does demonstrate that a distinction between genetic and environmental causes is not as clear as one might have thought.

The general conclusion of this and other examples is the following. The answer to what is *the cause*, or what are the causes, of a specific event has both an objective and a subjective, or pragmatic, aspect. The objective part of the answer is to give the necessary conditions of the complex that, as a whole, is sufficient for the effect.

The pragmatic, subjective or biased part of the answer is the proposal of one of these conditions as *the cause*, as opposed the other conditions in the complex, i.e. the background conditions. This selection is generally governed by our various interests.

7.2 Cause-Effect and Order in Time

In the discussion above we have looked at examples of both singular occurrences and connections between classes of events. In the first case there are causal connections between singular events, and in the second case causal connections obtain between classes of events. Singular events always occur within a specific time frame. Cause-effect relation is asymmetrical, that is, if event A causes event B, it is impossible for B to be the cause of A. The basis for this asymmetry is that cause must precede effect. In general it is the case that the cause precedes the effect for every individual cause-effect pair in a class of event pairs (There are cases where we are prone to say that two events are related as cause and effect albeit they occur simultaneously, or very nearly so. But this is a limiting case.). The definition of a cause as an inus-condition does not include this aspect of causation; rather, it is taken for granted.

Why does a cause typically precede its effect? The better question is perhaps 'why have we formed the concept of a cause, which operates in a manner that cause precedes effect?' The answer to this question is that causes are typically used to identify nodes in the flow of events where we can step in and perhaps modify the course of those events; we want to find the cause of an event in order to prevent or recreate a similar event. Naturally, we cannot prevent an event that has already taken place. When we want to know the cause of an event, we typically want to know what to do.

There is thus a pragmatic aspect to selecting the right cause, as we explained in the previous section. It is worth noting that this view is consistent with the idea that a cause and an effect could be practically simultaneous.

7.3 Causes and Statistical Correlations

How do we research causal connections? The ideal method would be the following: to isolate the object one wants to study and then to vary the presence and strength of certain factors that may influence the state of the object, that is, the different components of the various possible complexes of sufficient conditions. By varying one factor at a time, and by measuring the variation in the state of the object, one can obtain quantitative descriptions of how each component influences the object.

In most disciplines, however, this is not feasible. Instead, one has to make do with statistical investigations of classes of objects and events. The type of information one can acquire from such statistical measurements is essentially correlations between various factors that answers questions of the form 'is there some statistical connection between variables A and B, and how strong is it' (See Sect. 4.2).

Before delving further into this discussion, we should note that a correlation can only be attributed to a pair of *sets of events*, which we can describe using stochastic variables, whereas a cause-effect relation basically obtains between two single events, and only indirectly between types of events.

Despite their name, stochastic variables are actually functions, i.e., mappings from sets of events to sets of numbers. As an example, consider the throwing of dice. Suppose we call the stochastic variable K. For each throw of the dice, K receives a definite value, the number on the dice. That a stochastic variable receives a certain definite value can be considered an event. Thus, when discussing possible conclusions one can draw from correlations, in regards to causal connections, the question becomes which causal connections exist between the two individual events *variable X has the value* x_i and *variable Y has the value* y_i, given these event types comprise two correlated variables.

Now suppose that there is a strong correlation between two variables A and B.[2] Does this mean that A-events cause B-events? No, of course not. For all we know, B-events could be the cause of A-events, since the statistics give us no way of knowing the temporal order of the events. There is also a third option, that A-events and B-events are caused by some unknown type of events, call them C-events, without there being any causal link between A-events and B-events. If two events have a common cause, it is likely that they will appear together more often than could be explained by mere coincidence; and so looking at sequences of the same type of events, they will be correlated. Hence if all we know is that A and B are statistically correlated types of events, no stronger conclusion than the following can be made:

- A is a cause of B, or
- B is a cause of A, or
- A and B have a common cause.

This statement, that every statistical correlation is based on one of these three types of causal connection, is called *Reichenbach's principle*, after the philosopher Hans Reichenbach (1891–1956).

A common objection to this principle is that the correlation in question might be purely accidental. This objection is valid only if by 'correlation' one refers the observed correlation of a sample test. Indeed, it is clear that one can find a correlation between two variables in a sample without there being any such

[2] Instead of talking about two stochastic variables, statisticians use the notion of a two-dimensional stochastic variable.

correlation in the entire population. As described in Sect. 3.6, it is not possible to draw any conclusion about the probability for an hypothesis, in this case the hypothesis that there is a correlation in the population, using only results of observation of a sample; either one has to start with a prior probability and uses Bayes' theorem, or determine the likelihood, i.e., the probability for the observed result, conditional on the hypothesis that there is a real correlation.

Then, if the probability for obtaining a correlation in the sample test is less than 5 %, conditional on the assumption that there is no correlation in the entire population, and one still obtains a correlation in the sample test, then one has obtained a significant result at the 5 % level. This result may be interpreted as showing that there is reason to believe that the hypothesis that there is a correlation in the entire population. But we haven't got any probability measure for this hypothesis.

In what follows, the word 'correlation' refers to correlations in reality, that is, in the entire population and not only in samples. It will be assumed that one has observed a correlation in a sample test, performed the significance test, and *correctly* concluded that there really exists a correlation in the entire studied population. Given this meaning of correlation, the objection above is invalid. This is clearly shown by the following thought experiment. Suppose we have chosen 10 objects out an unlimited supply of a certain type of object, and have subsequently observed two properties. Furthermore, we find that there is a correlation between the variables we use for measuring these properties. Of course, this could be a coincidence. Thus we perform a larger sample test, say of 100 objects, and find the same correlation, or nearly so. Of course, this could also be a coincidence, even if the probability that it is has diminished. We can continue this process ad infinitum. If the correlation obtains regardless of how large of a sample test we perform (which we can never know, but a sequence of larger and larger samplings showing a convergence to a certain correlation coefficient, may give strong support for the hypothesis that the real value is close to the convergent result), then it is, by definition, not a coincidence that the correlation obtains, provided the sequence of samples is randomly generated.

In order to determine which of the three alternatives in Reichenbach's principle is the case, we must acquire further information. One can often dismiss one of the alternatives from information about the temporal order: If the singular event B precedes the singular event A, then A cannot have caused B. But how does one determine whether A causes B, or if there is a third background factor that is the cause of both A and B? In some cases the background information is sufficient, as the following examples illustrate.

Example 1 An investigation conducted in West Germany during the years 1966–1980 showed a correlation between birth rates and the presence of storks; as the number of storks decreased, so did the number of babies. The correlation was so strong that the probability for a mere coincidence was calculated to be less than 0.1 %. Hence there are very good reason to believe that it is real correlation. But since we do not believe that storks deliver babies, or that new-born babies attract

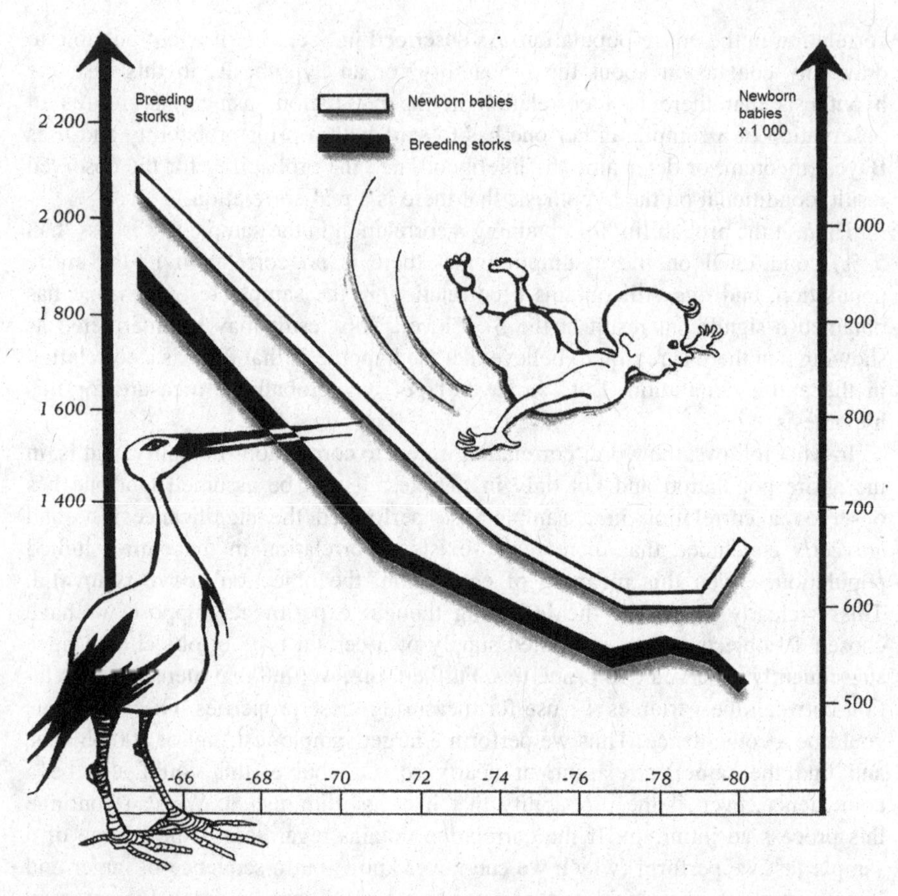

Fig. 7.2 Breeding storks and new-born babies in West Germany 1965–1980. The likelihood for the two variables not being correlated is less than 0.001 (*Source*: Nature, 7 April 1988. Reproduced with kind permission of Lou-Lou Pettersson)

storks, there must have been a hidden variable, a common cause. In my view, the most reasonable explanation is that both the decrease in the number of storks and the decrease in the number of babies are both the effects of an increase in German urbanization and industrialization during the period of observation (Fig. 7.2).

Example 2 An investigation of a cross-section of a certain population revealed a correlation between intelligence and shoe size. The bigger the feet, the more likely it was that one would perform better than the majority on an intelligence test. Is there in fact a correlation between foot size and the abilities of the brain? No, it is likelier that some background factor is at work here, namely age. Included in a cross-section of a population are a number of children of various ages, which have smaller feet and less likely to score well on an intelligence test than adults. This explanation can be easily tested by sorting out all children under the age of 18, and then checking if the correlation stills obtains. If it does not, then the correlation was the result of the age factor.

Assume that we do not have any background information that would help us determine whether a correlation between X-events and Y-events depends on a causal connection between them, or whether they have a common cause. If an experiment is feasible, the simplest test is to vary the incidence of X (where X precedes Y) and observe whether the incidence of Y follows suit. If so, there is an argument for X being the cause of Y. If the incidence of Y does not vary with the incidence of X, then X cannot be the cause of Y, which indicates that both X and Y share a common underlying cause.

However, it is often not practically, or ethically, feasible to perform the required experiment. In such situations one has no other choice but to calculate various conditional probabilities and observe which correlations obtain. In order to precisely say what conclusions can be drawn from information regarding correlations we need to understand two statistical concepts, viz., *statistical independence* and *conditional probability*.

Def. of Independent events: Two events A and B are independent of one another if and only if the probability that both A and B occur is the product of the individual probabilities, i.e. P(A and B) = P(A)P(B).

If this obtains, we say that the probability of the combined event, A and B, is *factorable*.

Def. of *Conditional Probability*:

$$P(A|B) = \frac{P(AB)}{P(B)},$$

where P(AB) stands for P(A and B). We express this equation as follows: the conditional probability of A given B is the probability of A in a population where B has occurred. The difference between the probabilities P(A) and $P(A|B)$ is that they are calculated in two different populations.

If two events, A and B, are independent of each other, it follows that the coefficient of correlation $r_{AB} = 0$ (see Sect. 4.2) i.e. they are not correlated. By Reichenbach's Principle, neither A nor B can cause the other. On the other hand, if A causes B (or B causes A), then a correlation exists which leads to A and B being dependent, i.e. $P(AB) \neq P(A)P(B)$. The converse, however, is not valid. Two events can be uncorrelated while still being statistically dependent.

(Note the difference between converse and contraposition. Given a proposition of the form 'if A, then B', we can form the converse 'if B, then A'. However, the contraposition to 'if A, then B' is 'if not B, then not A'. The contraposition of a proposition, as opposed to the converse, is a logical consequence of the original proposition.)

Now suppose that we have a correlation between variables X and Y and we know that X-events always precede Y-events. At this point we ask ourselves, are the X-events one of the causes of the Y-events, or do X and Y-events share a common cause? Suppose that we believe Z to be this common cause. Using conditional probabilities we calculate P(XY|Z), P(X|Z) and P(Y|Z) and we get either

$$P(XY|Z) = P(X|Z) \cdot P(Y|Z)$$

or

$$P(XY|Z) \neq P(X|Z) \cdot P(Y|Z)$$

(In practice, of course, we will never get an exact equality; if P(XY|Z) is close to P(X|Z) · P(Y|Z), we have to make a decision: equal or not?) Assume that P(XY|Z) is factorable, as in the first alternative above. What conclusions can be drawn? One possibility is that Z actually is a common factor of X-events and Y-events. Another possibility is that the factor Z stands for an event in between an individual X and Y-event (Note that even in this case P(XY|Z) is factorable.). If this is case, it follows that X-events are indirect causes of Y-events, a possibility that cannot be excluded.

If it so happens that P(XY|Z) is not factorable, as in the second alternative above, then Z is neither a common cause of X and Y, nor an intermediate factor. This result does not exclude the possibility of some other common cause of X and Y. All it means is that we cannot know for sure that X is a cause of Y.

Unfortunately, it seems very difficult draw any definite conclusions about causal connections from information about statistical correlations only, even if that information is significant and we have correctly concluded that these correlation obtain in the entire population (i.e., we have not made an error of the first kind.) More information is needed. Nancy Cartwright, using arguments different from those presented here, has come to the same conclusion, formulated as 'No causes in, no causes out' in her (1989).

7.4 Risk Factors and Conditional Probabilities

Risk factors for various diseases are often discussed in medicine. Fore example, smoking, high cholesterol and a certain genetic disposition are considered to be factors that increase the risk of a heart attack. This means that the probability of having a heart attack is increased if any of these factors are present. Thus the conditional probability of a smoker having a heart attack is higher than the non-conditional probability. This means that the number of heart attacks in a group of smokers is higher than in the entire population, which can be expressed as

$$P(\text{heart attack}|\text{smoking}) > P(\text{heart attack}), \text{or } P(H|S) > P(H).$$

Applying the definition of conditional probability,

$$P(H|S) = \frac{P(HS)}{P(S)}$$

we get

$$\frac{P(HS)}{P(S)} > P(H)$$

Rearranging the terms, we get $P(H|S) > P(H) \cdot P(S)$. H and S are evidently statistically dependent, which implies that there is a correlation between smoking and heart attacks (In this case there are only two values for each of the two variables, i.e. smoker/non-smoker and heart attack/no heart attack, which means that the correlation cannot be calculated in terms of the product–moment presented in Sect. 4.2; but a different form of correlation can be calculated.). My previous conclusion, that a correlation should not be automatically interpreted as a causal connection, also applies to risk factors. *A risk factor need not be a cause*. Rather, it could be a side effect of some circumstance, which both causes the disease, and, in another causal chain, causes the risk factor. In such cases, risk factors function as warning flags. It follows that if such is the case, there is no point in trying to decrease or eradicate the risk factor in order to decrease the impact of the disease. Doing so would be like turning off a warning light in order to avoid the catastrophe that the light warns us about. It is therefore essential to determine whether or not a risk factor is a cause (But in this case it is well established that smoking is an inus-condition, and can thus be said to be a cause, of lung cancer.).

However, it is worth pointing our that one does not often use the term 'risk factor' for factors that we believe not to be causal, even if they correlate with the disease. Indeed, there seems to be a shift in use of the word 'risk factor' from meaning 'statistically relevant factor' to 'statistically relevant factor, which comprises a part of the total cause'. This unclarity is deplorable and Olli Miettinen at Harvard suggests a remedy:

> Determinants of incidence are commonly referred to as risk factors. This term is a misnomer. Since the relation of an occurrence parameter to a determinant need not be the result of a causal connection, and since the term 'factor' (from the Latin word for doer) suggests causality, 'risk factor' is not a proper substitute for 'determinant of risk' A proper synonym is 'risk indicator' – analogously with 'economic indicator', 'health indicator' and so on.[3]

[3] Miettinen, O. (1985). *Theoretical epidemiology* (p. 10). New York: Wiley.

7.5 Direct and Indirect Causes

If A causes B, and B causes C, then A is an indirect cause of C. An indirect cause is still a cause, according to the above definition, since if B is a necessary part of a complex of sufficient conditions for C, and A is a necessary part of sufficient conditions for B, then A is also a necessary part of sufficient conditions for C. Thus A is a cause of C. The causal relation is hence transitive, which means that we can talk about causal chains. Every causal description is, therefore, infinite in principle. It also means that when we say that some event A causes another event B, there can, in principle, be infinitely many intermediate events. Thus the cause of some effect can occur long before that effect. In medicine there are many such examples, e.g. skin cancer, which can be caused by too much exposure to the sun decades before the disease arises.

Likewise, historians may discuss distant causes of important historical events. A. J. P. Taylor argues (in his *War by Time-table*) that the 'the man who pulled the trigger' and caused the first world war (1914–1918) was count Alfred von Schlieffen, German chief of general staff 1891–1906. How could he have 'pulled the trigger', he retired 8 years before WW1 began?

Schlieffen had construed a mobilisation plan for the German army on the presumption that France and Russia were allies and if war broke out Russia would take much longer time to mobilize than France. In case of war, therefore, the German army should first fight France with most of its force and then move to the eastern front. Hence, according to the plan, immediately after mobilization the major part of the German army was transported by train to Aachen, near the Belgian border. There is no place for millions of soldiers in Aachen so every division has, immediately after debarking the train, to march into Belgium in order to leave place for new troops coming. And since Belgium was neutral and England and France had promised to defend it, a German decision to mobilize was also in practice a decision to invade Belgium, which meant to start war. This was not fully realized by the Kaiser and his government; the German government's motive for deciding mobilization in August 1914 was to put political pressure on Russia, not to start war. So Taylor concludes that Schlieffen's mobilization plan was the cause of WW1 (And he shows, quite convincingly, that none of the great powers really wanted war.).

My own view is that Taylor has provided strong reasons to say that von Schlieffen's plan is an inus-condition for the first world war; and since no one really wanted war, according to Taylor, it is not unreasonable to say that the Schlieffen-plan was the cause (The causes of the first world war is a hot topic of debate, and by rehearsing Taylor's view I do not want to say he is right. But it is a prima facie plausible account of the causes and an interesting illustration to the notion of indirect causes.).

7.6 Causes as Physical Effects

According to David Hume, as we saw in Sect. 7.1, (i) a cause precedes, or at most occurs simultaneously with, its effect, (ii) causes and effects are always in close proximity, or connected by an intermediate chain of events also in close proximity, and (iii) a cause is regularly followed by its effect. One can characterize Hume's thesis as that he has given the necessary conditions for the use of the term 'cause'. Many philosophers have accepted Hume's thesis with the modification that a cause makes an effect more probable. But if we accept Reichenbach's principle, it follows that Hume's view cannot be correct, since the third condition, that a cause is regularly followed by an effect with a certain probability, is not able to distinguish between true cause-effect relations and common-cause correlations in which one factor happens to arise sooner than its correlated partner. Those who accept Reichenbach's principle thus require a deeper analysis of causation than what we get from Hume. Thus Reichenbach[4] and later Salmon[5] have further developed Hume's analysis of causation. Since Hume's second condition of proximity only applies to physical bodies, we need an analysis of causal connection that does not view links as made up of such objects.

A natural idea is to define a causal process as a transmission of energy (Prime example: the sun transmits energy to the earth and causes life.). That A causes B simply means that there is a physical connection, a signal of some kind carrying energy, which is transmitted from A to B. As far we know, there are only four fundamental kinds of interaction in nature: gravitation, electromagnetism, week and strong nuclear interaction. If an event A causes another event B, then these events must be connected by one of these four kinds of interaction, that is, there must be some exchange of particles doing the signalling from A to B. For example, in the human body, all connections between cause and effect, such as neural transmissions, are of an electric nature and neurons exchange charges. Mechanical interaction, such as friction or pressure, are also at bottom electromagnetic interactions: when two molecules collide with each other they exchange energy and momentum in the form of electromagnetic quanta, photons.

Since no signal can travel faster than the speed of light, causes must precede their effects no matter what coordinate system the observer uses. This fact agrees with our concept of a cause. If a cause and its effect appear to occur simultaneously, it is actually the case that the time it took for the interaction to take place is so small that it cannot be measured without sophisticated equipment.

From Mackie's analysis, it follows that a cause increases the probability that some effect will ensue. This is because, in most cases, a cause is part of a complex of necessary conditions, which must all be met for the effect to take place. Thus the

[4] Reichenbach, H. (1956). *The direction of time*. University of California Press.

[5] Salmon, W. (1984). *Scientific explanation and the causal structure of the world*. Princeton University Press.

coupling of a cause and an effect is only occasionally strictly deterministic, viz., in cases where there is only one inus-condition.

7.7 Cause and Effect in History

When studying history one is often interested in causes. For example, one might ask what caused the fall of the Roman Empire, the disintegration of Yugoslavia, or the industrial revolution. Can one apply the analysis given above to answer these types of questions?

The answer is yes. Mackie's analysis in terms of inus-conditions fits quite well in these situations, since we usually assume that historical events depend upon a large number of interacting causes. This analysis tells us that it is incorrect to ask for only one cause. Rather, we should try to identify all, or most, of the necessary factors, which together give rise to an historic event.

However, the next step in the analysis of causation, reference to physical mechanisms, seems, at least at first, more problematic. Is it really legitimate to talk about physical interactions as the basis of causal connections between historical events? It does not seem reasonable, but such is the case. Historical chains of events are comprised of a complex net of natural events (climate change, disease, famine, etc.) and the conscious actions of humans. It is obvious that the natural events are ultimately connected by physical mechanisms, but what about conscious actions?

Actions have both an internal side (thoughts, feelings and opinions) and an external side (our behaviour). Our behaviour, like natural events, is comprised of physical (chemical and biological) events, which are thus connected to each other via one of the four fundamental kinds of interaction. But what about our thoughts and feelings? In fact, even these can be understood as having a physical description (see Sect. 12.6). According to this view, historical events, including the historical agent's conscious considerations and unconscious motives, can be viewed as connected via causal chains (which are physical in nature) to other events, since they are at bottom physical states. In any case, there is no conflict between the idea that all causal connections are achieved via one or more of the four fundamental interactions and the use of the concept of a cause to link historical events.

However, it should be stressed that this fact implies nothing about whether or not historical events *should* be described as chains of physical mechanisms. The argument is not methodological but ontological. In history the physical aspects of events are not often of much interest. My point is merely to show that it is possible to give a single unified analysis of cause and effect that applies equally to the natural sciences, social sciences and the humanities.

Counterfactual Accounts of Causation

Causation has been a heated topic of debate among philosophers for decades. One particularly popular idea is to analyse the causal relation using counterfactual formulations. The general schema is something like

'A causes B' means 'if A had not occurred, B would not have occurred'.

This appears intuitively plausible, but I'm inclined to respond that I understand the meaning of the word 'cause' much better than the meaning of counterfactual propositions. So those who argue that counterfactual analyses of causation need a viable account of the semantics of counterfactuals. Many have relied upon the notion of possible worlds, utilized in the semantics of necessity and possibility. Roughly, the idea is as follows: in our world A occurs and is followed by B, whereas in a possible world, fulfilling certain conditions, A does not occur and neither does B. The crucial problem is, of course, how to identify those possible worlds where neither A nor B occur without using intuitions about causation, which in my view seems difficult. In fact, much of the discussion has been about purported counterexamples built upon our intuitions about causation. In other words, our intuitions about causation are used as adequacy conditions for a counterfactual analysis. But, why, then bother about counterfactuals, if we have stronger semantic intuitions about the concept of cause than that of counterfactuals? If anything, it seems more reasonable to analyse counterfactuals in term of causes.

An alternative account of counterfactuals builds on Judea Pearl's notion of structural equations, see Pearl (2000). The structural equations in a model of a type of system describe the dynamical evolution of the system being modelled. There is a structural equation for each variable of the form

$$Y = f(X_1, \ldots X_n)$$

where $X_1, \ldots\ X_n$ are the variables in the model. This equation is read by e.g. Woodward (2003) and Hitchcock (2001) as

If the variable X_1 would have the value x_1, X_2 the value x_2. ...., then Y would have the value $f(x_1, \ldots x_n)$

How do we obtain knowledge about such equations? By doing experiments and observations and generalise from limited number of observations to these general statements. In other words, causation is in this approach related to our possibilities to manipulate systems. In this I wholly agree, most often are we interested in causes because we want to know what to do in a particular situation.

Repeated manipulations of a system thus may enable us to formulate structural equations that express observed regularities. This is thoroughly discussed in Pearl's (2000). What this means, basically, is that this is a version of Hume's regularity account of causation, with the amendment that it is observed regularities which are brought about by manipulations.

7.8 Summary

The discussion of the concept of cause, as it is practically used, has led to the concept of inus-conditions. In cases where we want to pinpoint some such factor as the central or most important cause, this selection was shown to be a pragmatic one. We even found that the way we use the concept of a cause requires that the cause precede its effect and that both must be in close proximity, or otherwise connected by intermediate causes.

Cause-effect relations give rise to statistical correlations. What we observe are correlations, but if two events are correlated we cannot validly infer that they are causally related, because two events that share a common cause are also correlated. An important task in many scientific disciplines is to determine which of the three possible causal connections, listed by Reichenbach, accounts for a particular observed correlation. In fact, this can only occasionally be done using *only* statistical methods. Thus if at all feasible, performing the right experiment is a significantly more effective approach.

Finally, we placed the concept of a cause in our scientific world-view by conceiving of the connection between cause and effect as physical in nature i.e. based on one of the four fundamental kinds of interaction found in nature.

Exercises

1. In the 1940s, a strong correlation was found between the consumption of sweets and the frequency of caries among children. Could we conclude from this that a similar investigation would yield a similar result today?
2. So far, all attempts to establish a statistical correlation between so-called 'oral galvanism' (a taste of metal in the mouth, pain, headaches etc.) and the presence of other galvanic currents in the mouth have been unsuccessful. Does this mean that we can exclude galvanic currents as the cause of oral galvanism?
3. The following information is known:

   (a) Large amounts of substance X is a risk factor for the disease Y.
   (b) Lowering the amount of X in a group of people with high values of X in their bodies, but who have not yet contracted Y, does not reduce the risk for Y in this group.

   What conclusion(s) can one draw from these observations?
4. Suppose that a positive correlation has been found between low levels of a substance F in the blood and the incidence of the disease S. Which conclusion or conclusions can one draw from this?

   (a) Low levels of F is a risk factor for S.
   (b) Treating a group of patients with a medicine that increase the level of F in the blood will reduce the incidence of S.
   (c) Either the lack of F contributes to S, or S contributes to the lack of F.
   (d) There is a hidden factor causing both the low levels of F and the incidence of S, though the level of F in the blood has no effect on the incidence of S.

(e) None of the above conclusions can be drawn from the given information.

5. Given that the presence of factor F in the blood exceeding a certain limit causes disease G, which of the following is true?

   (a) If no amount of F is present in the blood, then one will not contract G.
   (b) If the amount of F in the blood exceeds the limit, one will contract disease G.
   (c) If the amount of F in the blood exceeds the limit, then the risk of contracting G is increased.
   (d) If one has G, then the amount of F present in the blood must have exceeded the limit.
   (e) If the amount of F exceeds the limit, the risk of contracting G will decrease if a medicine is administered to lower the amount of F in the blood.

6. The following example is fictitious. Out of a population of Swedish students, 31 % decide to seek higher education. Among students who have parents with academic training, 70 % seek higher education. Does this higher probability for higher education depend upon cultural factors, or does it depend on the fact that more educated parents have higher incomes? In order to answer this question, the following investigation was undertaken. The families whose income was 50 % higher than average were sorted out. Out of this group of families, 58 % of students went on to higher education. Furthermore, from this high-income group, 72 % of the students who chose to go on to higher education had academically trained parents. The families that contained at least one academically trained parent made up 80 % of the high-income group.

   What does this statistical data show? Is it higher education or higher income that causes a child from an academically trained family to seek higher education more often than other children?

7. Suppose you want to know whether there is a causal connection between two variables X and Y in a population. You select a sample from the population, measure the two variables and find no correlation. It is of course possible that there is a correlation in the population, so you make a statistical test. It shows that it is highly unlikely that there is a correlation in the population and you conclude there is not. Suppose this is true. Could one then infer that there is no causal connection between X and Y? Hint: Google 'Simpson's paradox'.

Further Reading

Box-Steffensmeier, J. M., Brady, H. E., & Collier, D. (Eds.). (2008). *The Oxford handbook of political methodology*. Oxford: Oxford University Press.

Cartwright, N. (1989). *Nature's capacities and their measurement*. Oxford: Clarendon.

Hesslow, G. (1984). What is a genetic disease. On the relative importance of causes. In Lindahl & Nordenfelt (1984).

Hitchcock, J. (2001). The intransitivity of causation revealed in equations and graphs. *Journal of Philosophy, 98*, 273–99.
Lindahl, B. I. B., & Nordenfelt, L. (Eds.). (1984). *Health, disease and causal explanation in medicine*. Dordrecht: Reidel.
Mackie, J. (1974). *The cement of universe*. Oxford: Clarendon.
Pearl, J. (2000). *Causality: Models, reasoning, inference*. Cambridge: Cambridge University Press.
Reichenbach, H., & Reichenbach, M. (1999[1956]). *The direction of time*. Mineola: Dover.
Salmon, W. (1984). *Scientific explanation and the causal structure of the world*. Princeton: Princeton University Press.
Salmon, W. (1998). *Causality and explanation*. Oxford: Oxford University Press.
Sousa, E., & Tooley, M. (Eds.). (1993). *Causation. Oxford readings in philosophy*. Oxford: Oxford University Press.
Woodward, J. (2003). *Making things happen: A theory of causal explanation*. Oxford: Oxford University Press.

Chapter 8
Explanations

'After the professor's explanation we were just as confused as before, only now on a higher level.'

8.1 Explanation and Prediction

The goal of science is a much-debated issue within philosophy of science. Is the goal of science to produce scientific explanations, or is it to make predictions? Some people attribute an essentially psychological meaning to the word 'explanation'; and hence infer that explanations are not important for the sciences. Instead, it is claimed that what is essential to science is accurate prediction. On the other hand, some say that the central aim of science is to satisfy our curiosity in regards to how the world works; to understand physical processes, social processes, man's history and people's reasons for their actions, which is acquired through explanation. Thus explanation is essential to science, whereas predictions only have to do with the application of science in technology. Technology is important, of course, but it is by-product of our understanding of how the world works.

A different view of this matter, held by Hempel, among others, is that the only difference between explanation and prediction is the temporal order of the act of explanation or prediction and the relevant event itself. Thus explanations regard past events, whereas predictions are about future events, but the relation between the *explanans* (that which explains/predicts) and the *explanandum* (that which is to be explained/predicted) is the same. Therefore, according to this view there is no interesting difference between explanation and prediction. It follows that if one can predict future events, then one can thereby also give a good explanation of those events once they have occurred.

I don't think Hempel's view holds water. There are many examples where one can make good predictions, but those predictions are not considered good explanations of a certain phenomenon. One can also give examples of the reverse: situations where good explanations can be given, and yet no relevant predictions can be made on the basis of those explanations. As an example of the first case, consider the correlation between the motion of the moon and ocean tides. Many people have observed this correlation over millennia; particularly in England,

L.-G. Johansson, *Philosophy of Science for Scientists*, Springer Undergraduate Texts in Philosophy, DOI 10.1007/978-3-319-26551-3_8

where tidal ranges are large and important for navigation at sea ports. High tide follows only a few hours after the moon travels past the meridian. Given that one has been able to rather accurately calculate the moon's motion ever since the time of the Babylonians, it has long been possible to predict the tides with reasonable accuracy. However, there was no explanation for this correlation until Newton published his theory of gravity. The fact is that Newton's ability to explain the change in tides was considered a monumental step forward for science, even though one had previously been able to accurately predict high and low tides.

An example of the second case, that one can give a good scientific explanation of some phenomenon without the possibility of making a prediction, is found within biology. Many biological phenomena can be explained with the help of the two basic principles of evolution,[1] but these principles cannot be used to predict anything. For example, consider the long neck of the giraffe. It is not difficult to explain why giraffes have such long necks given that (i) they live in the savannah, and (ii) gene variation occurs randomly. The giraffes that, due to random genetic variation, happened to have longer necks than other giraffes could eat leaves higher up in the trees and hence were better fit to reproduce. After a number of generations, the entire population came to have long necks. However, from this mechanism one cannot derive any prediction of which traits giraffes will acquire in the future, since the mutation of giraffe genes occurs at random and the disadvantages of a very long neck may come to outweigh the advantages at some point in the future via habitat change or some other mechanism.

In what follows, I will take for granted that explanations are different from predictions, even though predictions are sometimes facilitated by scientific explanation.

8.2 What Is Explained?

In everyday situations the word 'explanation' is used in many different contexts. In university education, as in other formal educations, new concepts are introduced and explained. This can be done in different ways, such as by giving explicit definitions, or by giving examples that show how a certain concept is used. However, this is not the meaning of explanation as it is discussed in the philosophy of science. Rather, what we are interested in is scientific explanation.

A scientific explanation can be roughly characterized as a line of reasoning that satisfies our curiosity about a phenomenon. In fact, much research is driven by the researcher's desire to understand mysterious phenomena. According to this view, which I share, explanation is the goal of all scientific disciplines. The question that

[1] These principles are the following: (i) individuals of the same species struggle for survival and the chance to reproduce, and those best adapted to their environment produce more offspring, and (ii) genetic variation in a species occurs randomly.

then arises is whether explanations have the same form in every scientific discipline.

According to a well-known position formulated by Dilthey and Weber, among others, there is an essential difference between explanation in the natural sciences and explanation in the humanities/social sciences (German *Kulturwissenschaften*). According to Dilthey and Weber, explanations in the natural sciences aim at producing descriptions of causal mechanisms, whereas in the cultural sciences, explanations aim to give *understanding*. The concept of understanding, as it is used here, contains an element of empathy (German *Einfühlung*). The objects of such explanation are actions and the artefacts and institutions those actions produce; these actions and artefacts have an intentional component that require interpretation, and are therefore not suitable for causal explanation.

Contrary to Dilthey's and Weber's views, Carl Hempel[2] claimed that all explanation, regardless of domain, has a common structure. This thesis has been much debated.

Another overarching issue regarding explanation is whether it is possible to determine the correctness or force of an explanation independent of the context in which it is provided. Many philosophers take it for granted, almost unconsciously, that the explanatory force of a scientific theory is a context-independent property that can be determined objectively once we find the correct general structure of explanation. In contrast to this view, a few philosophers, most notably Bas van Fraassen, have claimed that explanatory force is not something that a theory contains per se. According to this view, explanations are fundamentally dependent on their context; they are answers to questions people ask.

I will start by discussing the most well known model of explanation, the so-called Deductive-Nomological model (D-N model), developed by Hempel and Oppenheim, which claim universal applicability. Then I shall take up its statistical counterpart, the Inductive-Statistical model of explanation, which will be followed by a number of modifications of the D-N model. Following this discussion, I will present models for explanation that do not claim universal applicability, ending finally with a discussion of Bas van Fraassen's pragmatic model of explanation. This list does not exhaust all the various models of explanation. The debate surrounding this issue is quite lively, and as a result many ideas have been discussed. This shows that the concept of scientific explanation is multifaceted. The discussion will continue in the next chapter with various aspects of explanation that are especially relevant to the humanities and social sciences.

In the forthcoming discussion, two technical concepts will be important: 'that which explains', or *explanans* and 'that is which is explained', or *explanandum*.

[2] Carl Hempel (1905–1998) was born in Germany and worked in USA.

8.3 The D-N Model

The modern philosophical debate about explanation began with Hempel and Oppenheim's seminal 1948 paper *Studies in the Logic of Explanation*, in which they present the Deductive-Nomological model of explanation. The central idea of this model is that an explanation is a *deductive inference* from one or more *scientific laws* (law = $_{\text{Greek}}$ nomos) and *initial conditions*.

A classic example is Newton's explanation of why all bodies fall with the same acceleration, regardless of how heavy there are. According to an intuitive understanding of falling objects, it seems that the heavier an object is, the faster it falls, which was the general belief up until Galilei disproved it at the beginning of the seventeenth century.

Newton's explanation builds upon two fundamental laws; the law of gravity and Newton's second law, which states that a force (F) exerted on a body is equal to the body's mass (m) multiplied by its acceleration (a), i.e. F = ma. The law of gravity states that two bodies attract each other with a force proportional to the product of their masses and is inversely proportional to the square of the distance between them, as is given by the formula

$$F = G\frac{m_1 m_2}{r^2},$$

where G is the universal gravitational constant, which Newton calculated to be $6.67 * 10^{-11}$ Nm^2/kg^2 by experimenting with pendulums. If we treat falling as the result of the gravitational force between the falling object and the earth, we can equate the force due to gravity and the force causing acceleration, which gives us

$$m_1 a = G\frac{m_1 m_2}{r^2},$$

where m_1 is the mass of the falling body, m_2 is the mass of Earth, and r is the distance between the centres of the two bodies. This equation can be reduced to

$$a = G\frac{m_2}{r^2}.$$

Thus the acceleration of a body is determined by the universal gravitational constant, the earth's mass, and the distance between the falling object and the earth. Newton showed that the distance between two bodies, in this kind of situation, is the distance between their centres, and thus that we can use the earth's radius and a measure of r. The earth's mass is $6.0 * 10^{24}$ kg and its radius is $6.4 * 10^6$ m. Plugging these values into the formula above gives us the acceleration, namely 9.8 m/s^2. Thus we have a mathematical, i.e. logically valid, inference to the conclusion that free fall acceleration is constant and independent of the mass of the falling body.

Such an argument is, according to Hempel, a good scientific explanation of why free fall acceleration is constant for all bodies. The two essential elements of this explanation are (i) that the argument is a *logically valid inference* to the explanandum, and (ii) that the explanans contains at least one scientific law. Hempel presents these two properties as necessary and sufficient *formal* criteria for explanation. The only other requirement is that the propositions contained in a given explanation are true.

Schematically, we can organize the explanation of free fall acceleration in the following way:

Law 1:	$F = G\frac{m_1 m_2}{r^2}$
Law 2:	$F = ma$
Initial Conditions:	Earth's mass $= m_2 = 6.0{*}10^{24}$ kg
	Radius of Earth $= r = 6.4{*}10^{6}$ m
Conclusion:	a = 9.8 m/s^2

Hempel argued that all explanations must fulfil these two formal criteria, and conversely that all arguments that fulfil these criteria, and whose premises are true, are explanations. Thus he claimed that all explanations, if they are truly explanations, have this structure, regardless of whether they pertain to physics, biology, history or any other science. Before we examine whether these criteria are actually valid in sciences like history, let us discuss whether this model is correct within its primary domain of application, the natural sciences.

It is easy to find lots of examples of arguments within the natural sciences that we immediately understand to be good explanations, and that have the aforementioned structure, but it is just as easy to find counterexamples.

8.3.1 Problems with the D-N Model

(i) An explanation requires access to scientific laws. How have we come to know such laws? Well, quite often we have guessed or inferred them from the phenomena they are meant to explain. For example, one argument for believing the law of gravitation is the observed behaviour of bodies in free fall. But for an argument to count as an explanation, it is reasonable to expect acceptance of the explanans to be independent of the explanandum. If the explanans is to explain the explanandum, then our reasons for deeming the explanans to be true cannot involve the explanandum. Hempel's model thus requires that before the explanation is presented, we somehow have knowledge of the scientific laws involved and accept them as true. As we shall see in the next chapter, this is a big problem.

(ii) Explanations are asymmetric even as regards the laws. This asymmetry is missing in the D-N model. Consider the following explanation:

Law: If stormy weather is coming, the barometer falls.[3]
Initial Condition: Stormy weather is coming.

Explanandum: The barometer falls.

This seems to be ok. If someone asks why the barometer fell, one would give an explanation very similar to the above example. The argument is intuitively acceptable and fulfils the two criteria for explanation, but compare it to the following:

Law: If the barometer falls, then stormy weather is coming.
Observation: The barometer fell.

Conclusion: Stormy weather is coming.

This example also fulfils the requirements of the D-N model (for argument's sake, let's assume that the law 'If the barometer falls, then stormy weather is coming' is at least approximately correct), but one cannot very well accept it as a correct scientific explanation for why stormy weather ensues. That the barometer falls does not explain why there is stormy weather; rather, the opposite is the case: the stormy weather explains why the barometer fell. There is asymmetry between explanans and explanandum; that is, if A is part of the explanation for B, then B cannot be part of the explanation for A.

If we generalize the above example, we see that the requirement of asymmetry entails that the laws used in D-N explanations must also be asymmetric, in the sense that the reasons for believing 'If A, then B' are not good reasons for believing 'If B, then A'. This in turn shows that a correlation cannot suffice as a premise/law in an explanation. For if there is a strict correlation between A and B, then we can express it in the canonical form 'A if and only if B'. If this is true (in the case of partial correlations we can replace/generalise the expression 'If A, then B' with 'The probability of B, given A' and vice versa), then we do not have asymmetry. This shows the correlations lack explanatory force.

Furthermore, not all sentences of the form 'if A, then B' for which there is empirical support are accepted as laws. What is more needed than empirical support? This question is discussed in Chap. 10.

(iii) The D-N model allows for the explanation of a single fact from a general statement. This is normally not considered a real explanation. Here is one example:

Law: All ravens are black.
Initial Condition: The observed object is a raven.

Conclusion: The observed object is black.

[3] It is doubtful whether we really should call this a law, because (i) it is a qualitative statement, and (ii) there are exceptions. But Hempel and Oppenheim didn't mention any strictures of what to count as a law and their own example is similar in these respects. And the problem here under scrutiny doesn't depend on these aspects.

One can hardly accept this argument as an explanation for why the observed raven is black, since in such a case it is not the singular observation that requires explanation, but rather the fact that *all* ravens are black; we would normally not classify the observed object as a raven if it were not black. If someone were to observe a white raven, we would all want an explanation for why it is white and not black like all the rest, but in the case of observing another black raven, one would probably want to know why *all* ravens are black, and not just why the observed raven happens to be black.

This problem arises because of the assumption, held by proponents of the D-N model that it is primarily singular events we want explained. This is often not the case. We often require scientific explanations of repeatable phenomena, and not of unique events. (Of course, the opposite can be the case in regards to human action, but we shall discuss this later.) Thus the D-N model does not place sufficient requirements on what can be accepted as an explanation.

At least two ways of improving this model have been proposed. The first is to restrict the types of laws allowed in explanations to only *causal laws*. One would then be able to say that the D-N model is a model for causal explanations, which is what Hempel and Oppenheim probably thought, even though they made no distinction between causal laws and non-causal regularities. The other proposal is called 'unification'. The idea is that a phenomenon can be explained if one can show that it is logically associated with other previously unrelated phenomena.

8.4 Causal Explanations

One way to improve the D-N model is to restrict the type of laws required for explanations to only causal laws. With this requirement, one could say that the barometer example does not refute the model, since 'if the barometer falls, then stormy weather is coming' is not a causal law, and thus the barometer example is not an explanation. On the other hand, it is reasonable to claim that stormy weather causes the barometer to fall, that is, if one includes low air pressure in the concept 'stormy weather'.

We have already discussed the issue of how one distinguishes between causal laws and other laws in the previous chapter. We saw that all laws that relate variables to each other, whether they are deterministic or probabilistic, are correlations, and we have identified the conditions that must be satisfied for there being a causal connection between two variables. Alas, we are often unable to determine whether these requirements are fulfilled, but the point is that there is a distinction. Returning to the barometer example above, with the help of meteorological knowledge we can determine that the connection between stormy weather and a low barometer value is a correlation caused by a common factor; namely, low air pressure.

As we will see in Chap. 10, scientific laws have the logical form 'For all x, if Ax, then Bx'. Thus a precondition for a causal explanation is that 'Ax' describes a cause of the event Bx. (See the Appendix for a discussion of the logical form.)

In summary, one can say that a causal explanation is a logical inference from the cause to the effect based on a causal law.

8.5 Explanation as Unification

Another modification that has been proposed to improve the D-N model is *unification*. The central idea is that, though an explanandum is logically inferred from the explanans in agreement with the D-N model, the explanatory force of an argument does not primarily lie in this particular inference, but in the fact that many disparate phenomena can be similarly explained. This union is accomplished when the explanans not only explains a particular explanandum, but also others simultaneously. In other words, a proposed scientific law that only explains one type of empirical phenomena does not have much explanatory force. To illustrate this idea, once again consider Newton's law of gravitation. If it could be used only to infer free fall acceleration, it would not be much of an explanation of such phenomena. For a ponderous person could ask, 'what reasons do you have for believing in the law of gravitation?' The answer would then be that the law of gravitation explains the constant acceleration during free fall, which makes the reasoning circular. However, we know that the law of gravitation can be used to explain a great many seemingly unrelated phenomena for example:

- Kepler's three laws for planetary motion,
- The motion of pendulums,
- The tides,
- The orbits of comets,
- The earth's flattening at the poles,
- The precession of the vernal equinox

This means that our reasons for believing the law of gravitation are not limited to the particular phenomenon we want to explain in the first place, but that there are other independent empirical observations that serve as evidence for the law of gravitation. It is this property of the law of gravitation that gives it its explanatory force; the law of gravitation unifies the description of a number of phenomena into a coherent theory. An added advantage of this model is that it apparently contains different degrees of explanatory force, because it seems natural to talk about degrees of unification.

This view of explanation avoids the three previously mentioned problems with the D-N model: (1) we get an explanation of how we come to believe the explanatory law, (2) *unification* is asymmetric and (3) trivial generalizations have no explanatory force. However, it is not easy to formulate this idea into a concrete model. In this author's estimation the most natural model is that presented by

Michael Friedman.[4] Unfortunately, this model has been shown to have serious problems. Another variation, put forth by Philip Kitcher,[5] has attracted more interest. However, a deeper discussion of these theories' strengths and weaknesses falls outside of the scope of this book.

8.6 Statistical Explanations

Statistical explanations are explanations that contain probabilistic arguments. Hempel argued that there is a strong analogy between deductive-nomological and so-called *inductive-statistical* explanation. The difference is – according to Hempel – that instead of deterministic laws in the explanans, there are statistical laws. It follows that the explanandum cannot, in a strict sense, logically follow from the explanans. Rather, it follows with a certain probability.

Statistical laws have the form 'Event Y occurs with probability p, given that event X has occurred', which in mathematical notation becomes $P(Y/X) = p$. Thus the structure of explanation is schematically

$P(Y|X) = p$
X has occurred
------------------------p
Y occurs.

The statement that Y occurs is not a logical consequence of the premises, which is indicated by the dotted line.

The word 'event' should be understood in the mathematical-statistical sense: an object having a property is an event. According to Hempel, a statistical explanation is good if the probability p is close to 1. The model is called 'inductive' because we cannot logically deduce that Y will occur, even if X has occurred. But we do have statistical evidence to rely on, which can be formulated into statistical laws.

One central problem, which Hempel himself noticed and discussed, is that the probability of an event occurring seems to depend upon how we describe that event. An example is the following: Olga Svensson, 56, works as a public servant in Sweden. What is the probability that she will vote for the social democrats at the next election?

Given that a large number of middle aged women working in the public service sector in Sweden vote for social democrats, one might guess the probability to be rather high; say 60 %. If we now suppose that Olga is a doctor, and that it is widely known that doctors' political preference is more conservative, then this probability must be revised to a lower value. If we further suppose that Olga is a professor at a

[4] Friedman, M. (1974). Explanation and scientific understanding. *Journal of Philosophy, 71*, 5–19.
[5] Kitcher, P. (1981). Explanatory unification. *Philosophy of Science, 48*, 507–31.

medical institute, then one must further lower the probability that she will vote for social democrats, since this portion of doctors tend to be even more conservative. Finally, if we point out that she is a member of the social democratic party, then the probability changes radically in the opposite direction. So what is the actual probability that Olga votes for the social democrats?

This example illustrates the so-called *reference class problem*: the probability of some event depends on which reference class one thinks the event belongs to. The probability that Olga votes social democrat depends on how you describe her; that is, to which reference class she belongs. If we describe Olga as a member of the social democratic party, then we get a high probability, and therefore an explanation of why she voted social democrat. But if we describe her as a professor at a medical institute, we get a high probability that she will not vote social democrat, and thus an explanation of the complement event. This is not acceptable. One cannot reasonably say that one both has an explanation of some event and an explanation of the 'opposite' event.

Hempel proposed that one could solve this problem by requiring *maximal specificity* for the event to be explained. Maximal specificity is related to available knowledge. In the case above, the event X would be a complete description of all of Olga's *known* properties.

The problem with this solution is that the resulting inductive-statistical explanation is relative to what we know. Hence explanatory force cannot be an objective property of the theory. This conflicts with the fundamental assumption, which many philosophers – including Hempel – made early on, that scientific explanations must be objective and independent of the knowledge of observers. Given that explanatory force is not objective, Hempel was forced to give up the connection between explanatory force and truth.

It might be possible to avoid the relativism of explanation due to available knowledge by stating that maximal specificity includes descriptions of all properties that effect the probability of an event's occurrence, known or otherwise. Assume then that we have described a certain event completely; that is, all circumstances that influence the probability of an event's occurrence are taken into account. Is it not then likely that we will find that the probability of this event's occurrence is either zero or one, since we have all the relevant information necessary to make that determination? If so, then we no longer have a statistical explanation, but a deductive-nomological one. We thus may infer that there are no such things as objective inductive statistical explanations. Conversely, we should accept that where these explanations are employed, as they often are in the sciences, that the probabilities used in these explanations do not reflect objective, observer-independent facts.

The only way to sidestep this conclusion is to say that inductive-statistical explanations are only applicable to genuinely random events. A genuinely random event is such that even if we know everything there is to know about the conditions for an event's occurrence, we still cannot predict what will happen with certainty. I'll return to this topic in a moment.

Another problem is that, in many cases, we accept a probabilistic argument as a good explanation for events that are believed to be not genuinely random even though the event in question's probability is quite low. The connection between lung cancer and smoking is good example. Suppose a person has contracted lung cancer and that we ask for an explanation. We then view the information that the cancer victim smoked more than one pack of cigarettes per day for over 30 years as a good explanation. But, in fact, the probability of getting lung cancer if you smoke for over 30 years is only about 6 %. Still, we say that smoking is a good explanation for lung cancer. Therefore, Hempel's requirement that the probability should be high cannot be generally reasonable. Rather, it seems that the *increase* in probability when given new information is the essential explanatory step. The probability that a person will get lung cancer is very, very low, but if we add the information that a person has smoked for over 30 years, the probability increases many times over from this low baseline. It is this relative increase that we see as explanatory. This view suggests that what we take to be a good explanation depends upon what we know before the event, as well as what we find out from the explanation. Furthermore, this example and the entire discussion about inductive-statistical explanations also suggests that explanations have a contextual component that cannot be neglected. I will return to this point in Sect. 8.8.

I mentioned earlier that one could avoid the reference class problem in cases where the explanandum is a genuinely random event.[6] It is easy to see that this is the case. Suppose we describe an event in a certain way and give the probability of its occurrence. Suppose further that all relevant circumstances are contained in this description. This means that the addition of any other information will not change the probability. Thus the probability of the event's occurrence is independent of any further specifications, and a complete description of the event, if it were possible to give one, would also not change the probability. But then this probability must be the result of a genuinely random event, and is not due to incomplete knowledge of relevant circumstances.

Of course, one can never be certain that a probability is the result of genuine randomness. However, according to our best present theories about nature, there are some genuine random events; namely, irreversible state changes in quantum mechanics. The most well known example is the radioactive decay of atomic nuclei. In such cases one can, using quantum theory, determine the probability that an atom's nucleus will decay within a certain time interval. Furthermore, there is no additional knowledge that could change this probability. If quantum mechanics is a correct and complete theory, then we have a case of maximal

[6] It is not easy to exactly describe what is meant by a genuinely random event. Even most random events, e.g. computer generation of random numbers, can be shown to be determined by some complicated rule. One definition, due to Chaitin, is to say that a sequence is random if it is not possible to formulate an algorithm that produces the sequence and such that the algorithm is shorter than the sequence itself.

specificity, and the relevant probabilities that can be calculated using this theory are objective. Moreover, the explanatory force does not depend on the probability given.[7]

The conclusion seems to be that in situations where we cite a statistical argument as the explanation of some determined event without possessing all relevant information about the event, then one cannot give a strictly objective interpretation of the statistical explanation. That is to say, the explanation has an unavoidable contextual component, just as in the case of Olga's vote. Explanatory force is thus associated with what one knows prior to receiving the explanation, as well as any other information given when the explanans is presented. But in a case where the statistical law is an expression of genuine randomness and not a measure of our ignorance, one can give an objective and non-contextual analysis of the explanation.

8.7 Action Explanations

Human actions must be treated as a specific category when talking about explanations. This is because actions are intentional phenomena by nature. To explain an action is to view a human as an *agent*, whose actions can be judged in terms of what is rational or irrational. There are certainly exceptions to this rule, such as when one explains the actions of a psychotic person in terms of disturbances in the mechanisms of the brain. Yet we resort to this sort of causal-mechanical explanations only when the usual rational explanations, given in terms of beliefs and desires, fail.

We explain actions by giving the agent's reasons, i.e., citing the relevant motives or goals associated with those actions; this is more or less common sense. If we are unable to give an explanation in terms of motives, and instead attempt to give a biological explanation, we treat the action as behaviour. The difference is that when we describe something as behaviour instead of describing it as an action, we omit intentional components that we include, explicitly or implicitly, when we describe something as an action. Put another way, one can say that the concepts of *action* and *explanation in terms of motives* are logically connected.

We explain an action by citing its purpose or motive, as well as the agent's beliefs, which function as the link between the action and that action's goal. This means that the explanans contains two intentional elements: *goals* and *beliefs*. Thus explanations in terms of motives have the following form:

Agent A wants to achieve goal G
Agent A believes the best way to achieve G is to perform action H

Agent A performs action H

[7] Peter Railton has observed that the reference class problem can be avoided in cases where probabilities are interpreted as propensities, see Railton (1978). And since transition probabilities in quantum mechanics can be calculated without knowledge about frequencies, they can be viewed as propensities.

There is no guarantee that A in fact performs H, even if she wants to achieve her goal and believes that the best way to reach that goal is by performing H. For example, she could lack the necessary will, or have other goals that conflict with action H. Therefore, this explanation is not a logical inference of the explanandum from the explanans.

Suppose that we observe action H, or rather its behavioural component. In order to explain this action, we must first interpret it as a conscious action by ascribing an intention to perform the act to the agent concerned. We must then find out what beliefs and goals the agent has. However, this is not sufficient. For the explanation to be complete we must know that the action was performed for *just these reasons* and no others. It is not uncommon that the true reasons for an action differ from those explicitly stated, or from those that seem the most reasonable. Thus explanation inevitably contains elements of interpretation as regards the agent's actions and statements.

Notice also that the explanans contains an element of direction towards the future, toward the agent's desired state. Thus some have concluded that that which explains an action lies in the future. However, this view confuses the content of a desire with the mental act of having that desire. The act of having a desire occurs before, or perhaps simultaneously with, the action in question, though the content of this desire is identified by a description of a possible future state of affairs. That which explains an action is the fact that an agent has a certain motive, and this mental state occurs prior to the action. That the content of a motive is directed toward the future doesn't imply that the motive itself is held in the future. So the common argument that action explanations cannot be a species of causal explanations since causes precede their effects and that action explanations are future-oriented, is not valid: having a goal and a belief could cause an action though their content be about future states of affairs.

But there is another reason for not holding action explanations to be causal explanations: causal explanations require causal laws, and there seem to be no strict laws in the mental realm, only rough regularities, such as 'if one is successful in a difficult endeavour, then one becomes satisfied.'

Davidson has in his (1963) defended the common sense view that reasons, i.e., desires and beliefs, explains actions and that it is a species of causal explanation, without reliance of laws. Several philosophers have argued that Davidson is wrong on this point, holding that reasons, i.e., desires and beliefs, are not causes of actions. There has been an intense debate about this issue but no clear conclusion has generally been accepted.

In Chap. 1 I claimed that action explanations are, presumably, the original model for explanations of all phenomena. Natural phenomena were explained by the will of gods, spirits, or other conscious beings in a prime example of overgeneralization. It was a huge step forward when people realized that explanations in terms of motivations were not applicable beyond human actions.

8.8 Pragmatic Explanations

All the theories of explanation hitherto considered take for granted that the explanatory relation is only a relation between explanans and explanandum and that it is possible to determine whether an argument is an explanation without considering the context of the act being explained. Peter Achinstein has formulated this idea in the following way: when we have been convinced that the statements which make up the explanans are true, we should be able to determine, a priori, whether or not the purported explanation really is an explanation.[8] And Achinstein has argued – convincingly in my view – that for all the well-known models of explanation there are examples where this condition is not met. This indicates that the context of explanation has greater import than normally assumed. This is the basic idea in Bas van Fraassen's account of explanation, which is presented in his book *The Scientific Image*.

Van Fraassen claims that the question as to whether an explanation is correct can be answered only when it is given in the following form: is the given explanation good, or correct, *in this context*? The contextual component is made up of what the questioner knows or believes prior to looking for an explanation. According to van Fraassen, an explanation is *an answer to a why-question*. Thus the theory of explanations is a part of the theory of questions and answers.

Van Fraassen views the following as a typical explanatory situation:

Question Why P?
Answer P is the case, as opposed to Q, R, S, ..., because A.

P is called the *topic*, which is the state of affairs about which the questioner is inquiring. Furthermore, the questioner assumes a set of possible, although not actualized, alternative states {Q, R, S, ...}, which together with P are called the *contrast class*. In many cases these alternatives may be implicit. The why-question is thus formulated with the implicit assumption that P is the only actual state of affairs. Hence the first condition for answering the question 'why P?' is thus that 'P' is true and that all other alternatives in the contrast class are false. Suppose someone were to ask you 'why is the sky green?' You could not reasonably explain why the sky is green, and thus the only rational answer would be to reject the implicit claim that the sky is green. The second condition is that the answer, A, must also be true, and the third is that the answer must be *relevant* to the question. In other words, there must exist a *relevance relation* between the question and the answer. What this condition amounts to is not clearly expressed by van Fraassen. Two critics, Kitcher and Salmon, have shown that if one does not place any requirements on the relevance relation, one can construct a relevance relation such that any arbitrary true statement can answer any arbitrary question![9] Kitcher and Salmon drew the conclusion that one must impose certain requirements on the relevance relation.

[8] Achinstein (1981, 1983).

[9] Kitcher, P., & Salmon, W. (1987). Van Fraassen on explanation. *Journal of Philosophy, 84*, 315–330.

However, there is a question as to whether these critics have correctly understood van Fraassen's intention with his pragmatic model of explanation. Van Fraassen has not replied on this critique, but a defence is easily come by. A consequence of his pragmatic view of explanation is that the relevance relation between question and answer is a pragmatic aspect of the situation; that is, the answer must be deemed relevant *by the questioner in a given situation*. This means that one cannot say, in general, that a certain answer is relevant or irrelevant to a given question, but only that an answer can be relevant to some person in a certain situation while being irrelevant to another person in a different situation. There are therefore no criteria by which we can determine, independent of context, whether or not a proposed explanation is good. This means that it is not possible to place any universally applicable or context-independent requirements on the relevance relation. The question as to whether some explanation is good or not is incomplete. What is meaningful to ask is whether a certain explanation, in a given situation, is acceptable for a particular person. (The only non-contextual requirement on an explanation is that the statements that make up the explanans are true.) According to van Fraassen, there is no such thing as *the correct explanation*. This also implies that a theory's explanatory force is not something that can be used as evidence for the truth or believability of a theory. Explanations are thus uninteresting as regards epistemology, according to van Fraassen, which leads him to reject explanatory force as a relevant factor in comparing two competing theories.

The conclusion is that if van Fraassen is correct, explanations are irrelevant to scientific epistemology; and consequently, we should not devote time and effort to discuss them. According to van Fraassen, the central question in science and the philosophy of science is, 'which theories agree with the empirical data?' Whether a theory has strong or weak explanatory force is a question about what the scientific community considers to be reasonable or obvious. These aspects are not relevant to the question about whether scientific theories agree with empirical data.

Van Fraassen claims that this pragmatic theory solves two important problems affecting most theories of explanation. The first problem is that not all requests for explanation are reasonable. The solution to this problem is easily found in van Fraassen's theory, since a condition for it to be meaningful to answer the question 'why P?', and so reasonable to pose the question in the first instance, is that P must be true.

The other advantage of the pragmatic theory is, according to van Fraassen, that it accounts for the asymmetry of explanation. If A is a relevant answer to the question 'Why P?', then P is not normally a relevant answer to 'why A?'

Arguably, van Fraassen's pragmatic theory is quite successful as regards explanations in ordinary contexts. However, is it not reasonable to say that, in certain scientific situations, the contextual component is either minimal or non-existent? In other words, in a situation where a scientific explanation is requested, one expects an objective, context-independent, and generally acceptable explanation with respect to all relevant scientific theories. For example, one could claim that the explanation of some physical phenomenon should provide the objective physical causes of the phenomenon in question. Van Fraassen could perhaps reply here that the discussion of causation in the previous chapter showed precisely that the choice of a cause for some event is context-dependent, and that the only context-independent proposition one can make is in regard to what factors make up necessary conditions. In response, one could say that all language use is context-dependent, and thus this is nothing unique to explanation. Furthermore, all of these obvious necessary conditions show up in a given explanation as so-called *Ceteris Paribus clauses*, which will be discussed in Chap. 10.

8.9 Summary

The word 'explanation' is used in many different ways in various situations. All together, 'scientific explanation' concerns the explanation of phenomena, and not the explanation of the content of different concepts.

There are two fundamental conflicts in the debate over explanation. The first is whether there is a general model that applies to all kinds of scientific explanation. According to Hempel, the D-N model is such a model, which he takes to be generally applicable to all sciences. In response to his critics, who argue that this model does not fit historical explanations, Hempel states that those explanations found in history are not complete, they are merely sketches of explanations, which lack many details. Once these details are filled in, the explanation should fulfil the requirements of the D-N model. The counterargument to this response is the following: if we are to construct a model for scientific explanation, we must first know the actual nature of explanations in each discipline. We cannot simply try to fit explanations into a pre-established model.

The second fundamental dispute concerns whether the concept of scientific explanation contains a contextual component. Hempel, and many of his critics, have taken for granted that the concept of scientific explanation is a non-contextual concept, whereas Achinstein and van Fraassen have claimed that the contextual component is essential. The consequence of this latter view is that explanatory force of a theory cannot be used as an evaluative parameter, since it is relative to the background knowledge of those receiving the explanation. Contrary to this view, one could argue that there is a certain basic perspective shared by all active scientists in a given discipline, which is sufficient for the discussion of a theory's explanatory force independently of context, within that discipline. In van Fraassen's terminology, the relevance relation is decided by the theoretical perspective of the active researchers in a given discipline.

Exercises

Below are a number of examples of explanations, some of which are very sketchy. Discuss which model best fits each explanation.

1. Between 1850 and 1920, over one million Swedes immigrated to USA. Why? Historians claim that there were essentially two types of factors that explain this immigration. The first type is the push-factors, such as overpopulation, poverty,

the state church's oppression of 'free-church' movements, etc. The second type was the pull-factors; cultivation of the prairie gave many people the possibility of earning a decent living, leading early immigrants to write home about their success. Once emigration from Sweden was under way, the door was open for those who had family member or friends already settled in the U.S. to travel west and join them.

2. For more than 40 years, Pripps (a Swedish brewing company) had been licensed to produce Coca Cola in Sweden. In 1997, Coca Cola voided Pripps' contract and decided to build their own factory to serve the Scandinavian market. Why did Coca Cola cancel the contract? An explanation given by the press is that Coca Cola was unhappy with its market share in Sweden, since they had a smaller portion of the soft drink market in Sweden than in other comparable countries. This is because Pripps, who dominated the soft drink market at the time, competed with Coca Cola in regards to some similar products. Thus Pripps was not interested in increasing Coca Cola's market share, as this would result in lower sales of Pripps' products.
3. Why did a bloody war in Bosnia follow the fall of Yugoslavia? A not too unreasonable sketch of an explanation is the following. At the fall of communism, Slobodan Milosovic, who was then party chief of the Republic of Serbia, saw where change was headed and that he was in risk of losing power. He chose to appeal to Serbian-nationalist sentiments in order to acquire a new political platform. Milosovic was successful, and the result was that many people dreamed of a Greater Serbia as the successor of Yugoslavia. Since the Serbians practically had control over the Yugoslavian army, their leader chose to start a civil war with the knowledge that they had superior military resources on their side. The goal was to create a cohesive area in Bosnia devoid of Muslims and then join with the remainder of Yugoslavia, thereby building themselves a dream state for all Serbians.
4. Why are carrots orange? Answer: The colour of carrots comes from carotene molecules. All molecules contain electrons that are situated at difference energy levels. In a carotene molecule, the difference in energy between the highest level and the next highest level is about 3 eV. When light hits a carrot, the light ray is absorbed with a frequency that corresponds to the energy gap, according to the formula $E = hf$, where h is Planck's constant and f is the frequency. Blue light has just the right frequency corresponding to the energy gap so that it is absorbed. Thus when we look at a carrot, it looks orange, since the colour complement to the absorbed colour dominates the reflected light.

Further Reading

Achinstein, P. (1981). Can there be a model for explanation? *Theory and Decision, 13*, 201–227.
Achinstein, P. (1983). *The nature of explanation*. New York: Oxford University Press.

Cartwright, N. (1999). *The dappled world. A study of the boundaries of science*. Cambridge: Cambridge University Press.

Davidson, D. (1980 [1963]). Actions, reasons, and causes. In *Essays on actions and events* (pp. 3–20). New York: Clarendon Press.

Hempel, C. (1965). *Aspects of scientific explanation and other essays in the philosophy of science*. New York: Free Press.

Hillel-Ruben, D. (Ed.). (1993). *Explanation. Oxford readings in philosophy*. Oxford: Oxford University Press.

Pitt, W. (Ed.). (1988). *Theories of explanation*. Oxford: Oxford University Press.

Railton, P. (1978). A deductive-nomological model of probabilistic explanation. *Philosophy of Science, 45*, 206–226.

Salmon, W. (1984). *Scientific explanation and the causal structure of the world*. Princeton: Princeton University Press.

Salmon, W. (1998). *Causality and explanation*. Oxford: Oxford University Press.

van Fraassen, B. (1980). *The scientific image*. Oxford: Clarendon Press.

Chapter 9
Explanation in the Humanities and Social Sciences

'It is not the consciousness of men that determines their existence, but their social existence that determines their consciousness.'

K. Marx

9.1 Methodological Collectivism Versus Methodological Individualism

There are two possible perspectives one can adopt when discussing social phenomena. One can either discuss phenomena with respect to some specific point in time, or else discuss phenomena in terms of historical development. The former adopts what is known as a synchronic perspective, while the latter adopts what is known as a diachronic perspective. In either case, one is confronted with a basic methodological choice: should one explain states of affairs, institutions, events, etc., in terms of the agent's motives, beliefs, desires and actions, or should one reverse the explanatory direction by looking instead at objective social forces in order to explain agents' actions and thoughts? The first alternative is called *methodological individualism*, and the latter *methodological collectivism*. If one opts for the first alternative one naturally starts using some version of the action explanation model, while if one prefers the latter alternative, some sort of causal explanation will be the most relevant.

From a methodological point of view, both alternatives have their pros and cons. A disadvantage of methodological individualism is that agents are sometimes inaccessible (agents of historical events are in most cases dead) and thus unavailable for questioning, leaving us with only the historical artefacts and sources they have left behind. These artefacts are often incomplete and difficult to interpret. Furthermore, one is often suspicious that the explicitly recorded motives of an action are fictitious, or outright lies. In politics, as in everyday life, few people are completely honest about their beliefs, motives and desires. Another drawback is that one sometimes gets the feeling that the individual agent's motives and beliefs are partly a product of the society in which they live. Therefore, an agent's individual characteristics may not be all that interesting or relevant. For, as regards a historically interesting event, it is possible that the characteristics of the agents involved did not play a large role. Perhaps they simply happened to be at the right

L.-G. Johansson, *Philosophy of Science for Scientists*, Springer Undergraduate Texts in Philosophy, DOI 10.1007/978-3-319-26551-3_9

place at the right time to fill in the necessary roles for some event's occurrence (or – as it may seem in some cases – the wrong place at the wrong time). The suspicion that agent's motives and beliefs are products of the society they live in is the basic intuition in methodological collectivism.

If we instead adopt methodological collectivism, another difficulty arises. Suppose we want to explain a specific event or the existence of a certain trait in society with the help of social forces, which operate more or less independently of the individual agents involved. Naturally, one would then give a causal explanation. If such an explanation is to be successful, one needs to discover the general causal laws connecting various features of society. This has proved an extremely difficult task. The difficulty lies in precisely stating the conditions under which a type of social phenomenon will occur. In principle, one could accept that there exist historical and social laws that manifest themselves only once, but how could we know anything about such a law? Furthermore, in order to have a successful explanation, according to methodological collectivism, we need causal laws that connect social phenomena with the thoughts and beliefs of the various agents in a society: one needs causal laws of the type, 'under objective circumstances X, people tend to desire (believe, think) Y'. Once again, this has proven difficult, if not impossible. To answer this concern one might opt for an account of singular causation ('event A caused even B') where laws play no role, but then there is a problem in regards to what is meant by a 'cause' in such a context.

Generally, one is a methodological individualist or collectivist on the basis of ones view of human nature. Methodological collectivists tend to view humans as social creatures by nature, whose character traits, beliefs and desires are shaped by interacting with other people. One exponent of the collectivist view was Aristotle who wrote: 'Man is by nature a social animal; an individual who is unsocial naturally and not accidentally is either beneath our notice or more than human.'[1] Another is Marx who claimed that man's consciousness is determined by his relations to others. There is certainly some truth in these statements, but a really explanatory theory about the relations between individual minds and the surrounding social structure is lacking. The problem is conceptual; the term 'social structure' covers phenomena that somehow are constituted by individual minds, so one need to think of the relation between individual minds and society as internal: the relata are constituted by the relation and so cannot be described and identified without reference to the relation.

Methodological collectivists can criticise the other camp for only going half-way when using individual beliefs and perspectives as explanans for social phenomena; Such an explanation does not exhibit the more basic elements of history and society, but only the manifestation of social forces in the minds of individuals. The methodological individualist may reply that human beings are autonomous agents whose beliefs, thoughts, motives and desires cannot be fully reduced to external social factors.

[1] Aristotle: *Politics, 1253.*

Hence, explanations in social and human sciences differ profoundly from explanations in the natural sciences.

Many philosophers, for example Popper, have claimed that the only reasonable stance is methodological individualism, since the existence of impersonal forces not based on the thoughts and actions of individual people is pure mysticism. In my opinion, this argument does not sufficiently take into account the essential social nature of human beings. There is a known case where a girl has been kept alive in the cellar by her father but not really being interacted with or spoken to for several formative years. When she was found, she got the best possible care and linguistic training, but she never became a fluent speaker and she was mentally retarded. This is but one drastic example, but the general conclusion is supported by a wealth of evidence. Maturing into a person with a mind requires intense social interaction.

It is obvious that more than one person is required in order to create languages and norms. Methodological individualists do not deny this, but they do claim that social forces are the results of individual actions. Thus this discussion appears similar to the question as to what came first, the chicken or the egg.

According to the methodological collectivist, history should be cleansed of descriptions of individual actions, because such events are determined by external, impersonal forces. An explanation that consists of a description of a chain of events containing some individual actions is thus considered insufficiently penetrating. An illustrative example is the Marxist view on the creation and development of the capitalist system. Marx claimed that a historical explanation of economic development misses the point if it draws upon individual desires and actions, (Why is the nobility so rich? They are greedy!) since these things arise in social situations. Rather, it is the development of productive forces – the technical and organizational development of material production – that determines the economic-historical process. The methodological individualist replies that every such technical or organizational change is the result of individual decisions guided by the individual's desire to better his or her own life.

We can, perhaps, see here the contours of a solution to the methodological collectivist's problem of describing social forces in a way that does not involve the conscious actions of agents. The basic regularities that govern human action can perhaps be described in biological terms, such as the basic desire to live, reproduce, and improve one's living conditions. These are regularities among humans that can be explained without resorting to mysticism.

It is not unreasonable to say that the choice between methodological collectivism/individualism is really just a choice of the starting point for an explanation. Every explanation consists in presenting certain statements, an explanans, as an explanation for some explanandum. An inquisitive person might accept the overall explanation and yet request an explanation for some of the circumstances expressed in the explanans. Just like causal chains, explanation chains progress, and regress, indefinitely. In other words, there is no phenomenon such that it does not require an explanation. From this perspective, one could say that the methodological individualist finds it reasonable to make the beliefs and desires of individual agents the ultimate starting point of an explanation, whereas the collectivist is more interested in the social forces that brought those beliefs and desires into being. The choice between methodological collectivism and individualism is thus a choice concerning which questions one finds most interesting and which facts one finds so obvious that they do not require explanation. This conclusion points towards a pragmatic view of explanation like the one discussed at the end of the previous chapter. In short, the choice between methodological collectivism and individualism depends upon the type of question one wants answered.

9.2 Explanations of Historical Events

In the case of historical events, it is difficult to claim that similar events occur: every historical event is more or less unique. It is therefore doubtful that there are any historical laws similar to those used in describing natural phenomena. In historical explanations, individual events make up both the explanans and explanandum. So then, what kind of relations must obtain between two historical events such that the first event explains the second event?

Every historical event depends on the interaction of a great number of factors. Thus according to many historians, the most reasonable model for historical explanation is *narration*. To explain an historical event is to describe a number of these factors such that the explanandum appears natural as the next step in the story. This view seems reasonably intuitive and appears to coincide with how most historical explanations are presented. The links of the chain of events that build up the story are of different kinds. Some are the actions of individual agents, some social processes and others are natural phenomena. Adherents to the narrative model of explanation are often methodological individualists, although most would accept that the individual agent's feelings, motives and beliefs are influenced by the circumstances in which the agent lives. However, narration, i.e., story telling, leaves many things unaddressed. Historical narration is often allowed to omit those beliefs and thoughts that we commonly consider obvious or natural. As regards western medieval history, for example, one need not explain why agents tended to have strongly religious world-views. But one would require an explanation for why certain individuals were *extremely* pious or *notably* irreligious.

The philosopher Paul Ricoeur, who forcefully argued for narration as the model for historical explanation, has claimed that Hempel was right in insisting that if a description of a chain of events is to have any explanatory force, the connection between two events must not be a literally unique case. For, if we are to understand the former event as explaining the latter event, we must see these events as instances of types of events that regularly follow each other. That is to say, there must be, at least implicitly, a law-like connection between two events in order for explanation to be possible. The narration must be presented such that the explanandum follows rationally and naturally in the given circumstances, which means that one must describe the circumstances such that one grasps the generality of the situation.

Narration can have either an individualistic or collective character. If a chain of a narrative contains a description of the origin of some agent's desires or beliefs (e.g. strong nationalistic sentiments), then the methodological collectivist can explain this chain in terms of social forces. This seems reasonable as long as the beliefs or desires in question are not agent-specific. However, if the agent (s) involved have peculiar or unique beliefs or desires, then perhaps what we want to know is how this peculiarity came about, which would require a more individualistic perspective.

All narration is selective. We choose a number of aspects out of a multifaceted reality and include these in our story, while other aspects are left out. This selection is governed by what one believes about the audience's perspective, as is always the case in situations of communication. Thus we have yet another case for the pragmatic character of explanation.

The following example illustrates a number of the points made above. It is an article written by leading economist Jeffery Sachs and published in the Swedish newspaper Dagens Nyheter on March 23, 1998.

> What caused the Asian crisis during the fall 1997?
>
> There are those who blame Asian capitalism for the Asian crisis: a devastating mixture of corruption, nepotism, governmental interference and bad financial control. This explanation suits western banks just fine, as it relieves them of all responsibility. But if Asian capitalism is so bad, then why did those banks invest so much money in Asia to begin with?
>
> A more plausible explanation is that the investors panicked. Perhaps they were too euphoric in 1996, and already too pessimistic the year after. In the end, the result was a devastating panic that hurt both investors and the Asian economy.
>
> The immediate cause was clearly a sudden change in the flow of capital. In 1996, foreign investors placed 93 billion dollars into Indonesia, Malaysia, the Philippines, and Thailand. In 1997, they withdrew 7 billion dollars. This swing of 105 billion dollars corresponds to 11 per cent of the national income of these countries and has driven them into a deep recession.'

The structure of this explanation, or rather suggested explanation, is the following. The Asian crisis is explained as being caused by negative capital flow. A causal law is thus being implicitly applied here; namely, '[i]f a country's capital flow changes rapidly in a negative direction, then this country will experience an economic crisis'. This 'law' is not very precisely formulated, and is perhaps not as deterministic as is suggested above, but it is obvious that Sachs uses such a law in his argument. It is also clear that Sachs takes this law to be well known and generally accepted, since he does not bother to argue for it separately.

The triggering factor in this case is a sharp decline in the flow of capital. But why was there such a drastic change in flow in 1997? The reason is, according to Sachs, that the investors panicked. This part of the outline is thus an explanation of an action. To panic is to enter into a certain mental state. In this case, the investors believed they were loosing money in Asia and that the best way to avoid this loss was to pull out of the Asian market as fast as possible. Obviously, the investors' goal was to earn money and avoid any losses.

In summary, we have here a short narrative consisting mainly of two coupled explanations. The first, the explanation of the change in the flow of capital, is an action explanation. The second, an explanation of the effect of this change in capital flow, is a causal explanation. However, neither explanation is complete. The story only contains the elements the author believes to be necessary in order for the reader to follow his argument, nothing more. The pragmatic element that shows itself here is the choice of factors that are to be included, and those that are omitted, in the explanation.

9.3 Explanation of Social Phenomena

Explanations of social phenomena, especially social institutions, are controversial in two respects. One is the conflict between methodological individualism and methodological collectivism, and the other concerns what kinds of concepts should be used. The latter dispute is most clearly seen within psychology and psychiatry where one either treats humans as agents, in which case action explanations are appropriate, or one treats humans as organisms that exhibit certain behaviour, in which case the appropriate explanation will be the one that explains that behaviour. Recall that *action* and *behaviour* are different things. When we describe something as behaviour, we do not assume that it is accompanied by any conscious mental processes, (or it may be tacitly assumed that it is, but treated as irrelevant) whereas when describing an action, we do.

Are there any unconscious actions? Yes. For instance, an actor can honestly deny having a particular belief without anyone taking his words at face value; few nowadays deny that our minds contain unconscious elements.

One can generalize this conceptual observation to all of the social sciences. Many, perhaps most, of the relevant concepts in the social sciences have an *intentional* component that refers to conscious or unconscious attitudes or beliefs. Consider, for example, the concept *buying something*. This concept is connected to a host of juridical and economic institutions, such as contracts, money, and property. If I buy something, then I *own* it and I have the *right* to use it. Furthermore, I am *obligated* to pay for this object. If someone *stole* the object from me, I would have the right to ask for the help of certain authorities to *get it back*. Most often we do not consciously entertain these beliefs, but we are aware of these social facts. All these concepts describe social relations between people based on *collective intentions* that exist among the members of the society in which the object was purchased.

An event that clearly illustrates the intentional character of these juridical/economic relations is the Dutch colonists' purchase of Manhattan from the Native American community, the Lenape Indians, who used the island as hunting ground. According to the colonists, they *bought* the land from the Native Americans for 60 guilders. However, the Lenape Indians did not think of the interaction that way. On the home page http://delawaretribe.org/blog/2013/06/26/faqs/#Living their descendants write: 'Our ancestors were asked to sign treaties giving up the land, but they had no idea that they were actually selling land any more than you would think someone could sell air. The belief was that all land was put here by the Creator for use by his children, and that you should not be stingy with it.'

One might guess that the Lenape people viewed the transaction as an act of diplomacy, through which the colonists were expressing their peaceful intentions by giving them gifts. Hence, whether we should classify the handing over of 60 guilders from the colonists to the Lenape people as a purchase or as a gift depends on whose perspective we speak from. If we ask 'but was this transaction really a purchase or a gift', we are erroneously assuming that one can use these concepts without referring to the beliefs of the people involved.

These two controversies, methodological collectivism/individualism and causal/intentional explanations, are intimately connected without being identical. An adherent to methodological collectivism tends to neglect intentional components of actions, whereas the methodological individualist stresses them.

Thus the question is whether the correct model of explanation in the social sciences is some form of causal explanation, or some version of action explanation. The answer depends, to a great extent, upon how one perceives human beings. If one views humans as primarily autonomous individuals, it is natural to adopt methodological individualism. If, on the other hand, one views humans as primarily social creatures, which have few or no properties (except biological ones) independent of their society, then it is natural to adopt methodological collectivism.

Any given view of human beings is a mixture of factual and ideological convictions. Although it may in principle be possible to formulate the question about methodological collectivism/individualism as a question about what is the most scientifically interesting way to study social phenomena, ideological aspects cannot in practice be disregarded.

One might think it obvious that a sociologist would be a methodological collectivist, since in this discipline one is mostly interested in collective phenomena and not the specific properties or actions of individuals. However, the question of methodological collectivism versus individualism is a question about what explains what. Can one explain social phenomena via a description of the involved individuals' properties, or are the properties of individuals explained by the social environment in which they live?

No matter which type of explanation is correct, one might think that, since many relevant concepts in the social sciences have an intentional component, the only reasonable model is that of action explanation. However, the matter is not that simple. First, the intentional aspect of social concepts is a collective intentionality, and not an individual one. Since the goal of many sociological inquiries is to investigate the genesis, structure, and development of social institutions involving such intentional components, one cannot take collective intentions as primitive elements of explanations, as it is precisely these one wants explained. Secondly, one can claim – as many have done – that intentions (individual as well as collective) are a type of causes. It is therefore reasonable to use causal models even in the social sciences.

In my view, collective intentionality is a central concept in the social sciences. Furthermore, it is a theoretical concept, as one cannot observe collective intentions.

A difficult problem is the relation between collective and individual intentions. The simplest view, which states that collective intentions are nothing but the sum of individual intentions among the members of a group of people, is clearly inadequate, since individual intentions are heavily influenced by experiences of belonging to different groups.

Besides these two alternatives (causal and action explanations), there is a third alternative known as functional explanation.

9.4 Functional Explanations

If we turn our attention towards scientific practice, it seems we quite often find that there is a special type of explanation in the social sciences; namely, functional explanation. An explanation is called functional if it explains a social phenomenon in terms of an important, though often unintended, function that phenomenon performs in a society. If a habit or trait that is not inherited serves no function, then it is not reproduced, and soon disappears. That is to say, there must be a feedback mechanism that helps some habit or trait be reproduced from time to time. Thus in some sense, the existence of a phenomenon is explained by its positive function.

An example that is often discussed in illustrating this point is taken from Robert Merton. Merton studied a number of American companies and discovered a significant difference in conflict level between successful and less successful companies. The more successful companies had higher conflict levels. Merton interpreted this fact as the result of the existence of a feedback mechanism. That is to say, if a company tolerated conflicts and allowed different opinions to be expressed, then the organization would become more flexible and open for change. The more employees were allowed to express their opinions, the more information regarding the ever-changing environment would flow through the company, allowing the management to more effectively adapt to economic changes and increase the company's competitiveness. In short, conflicts have a positive function; thus benefiting the organization that allows for them.

A much-debated question is whether or not functional explanation is actually a new and distinct type of explanation. Jon Elster has answered in the negative, suggesting that functional explanation can be reduced to either action explanation (if one assumes an individualistic perspective) or causal explanation (if one assumes a collectivistic perspective). For, how does the feedback mechanism work exactly? That a social phenomenon fulfils a function means that (assuming methodological individualism) every individual involved is aware of what decisions are considered rational and that the sum of these decisions lead to a certain social phenomenon being reproduced. For example, a political party exists only so long as a number of people actively decide to be members of that party. Thus the correct type of explanation in regards to the existence of political parties (in the individualistic perspective) is action explanation. In other words, the existence of political parties, organizations, corporations, or other social phenomena is never an unintended side effect. On the other hand, if we assume that the individual agents' beliefs and desires are, in turn, determined by social situations those agents find themselves in, then an explanation with the agents' beliefs and desires as the explanans is not satisfactory. Rather, the explanans should be comprised of a description of the social context and causal connections responsible for these beliefs and desires. According to this perspective, the individual's beliefs and desires are caused by external circumstances, which means that the feedback mechanism must be described as a causal connection, reducing the functional explanation to a causal explanation. In the example taken from Merton's investigation one does not assume any conscious intent to streamline the organization behind the agents' actions. The feedback mechanism must then be a causal mechanism. Merton's explanation is thus a good fit for the causal model of explanation, given that we ask ourselves why conflict-ridden organizations exist and prosper.

Functional explanations are apparently inspired by evolutionary biology, where the surface structure of many explanations is to explain the existence of a certain trait of a species in terms of its role, or function, in the survival of that species. However, this is only a surface explanation, and requires deeper analysis. A more complete evolutionary explanation can be formulated in the following way. A random mutation in the genetic make-up of an individual may give it a new trait, or property, which increases its chances of survival and reproduction. Thus it is likely that this individual will produce more offspring than the other members of its species. This results in an increase of individuals with the same beneficial property in the next generation. After a certain number of generations, this property will have spread throughout the entire species. Thus we have a fundamentally causal explanation of the spread of a certain property or trait. The feedback mechanism is purely causal, and the only non-causal element in the explanation is the appearance of spontaneous mutations.

If we now compare this with possibly similar mechanisms in e.g. organizations, the question becomes, what produces the feedback mechanism? If such a mechanism exists, and is such that it works without people being conscious of it, then it must be a causal mechanism, as in biology. If, on the other hand, the feedback mechanism is a conscious activity, reflecting our behaviour so as to better achieve our goals, then what we have is an action explanation at the collective level. According to Elster, there is no such thing as *functional* explanation, distinct from causal explanation or action explanation; a conclusion that I find well grounded.

Without going any deeper into this discussion, one can clearly see that there is an intimate connection between the type of explanations one deems correct for the social sciences and the basic views one has of humans as social creatures.

9.5 Summary

The debate about explanations in social and human sciences is multifaceted. The fundamental dispute concerns deep metaphysical issues about the relations between individuals and society and how one conceives a human being, a person. The dividing line goes between those who conceive individual human beings as basically independent of society and other persons, (which seems to be implicit in the liberal political tradition), and those who rejects this view, holding that a human being is constituted by its social relations. At the methodological level this distinction becomes methodological individualism versus methodological collectivism.

The individualist view on man is the default option in the western intellectual tradition, but there are good arguments for the other view.

Exercises

First, discuss whether the following explanations are expressions of methodological collectivism or methodological individualism. Then discuss which model of explanation best fits the argument.

1. (Translated from J. Elster: Økonomi og historie, Pax forlag, Oslo, 1972, pp 61–62)

   The pre industrial society could best be described with the word 'stagnation'. The economy was trapped at a low level, in a vicious circle of mechanisms inhibiting growth. Technical development stood still because the investment level was so low that capital-consuming investments never left the ground. This was partly caused by the fact that productivity was so low that there was rarely any surplus left for investment, and partly by the lack of interest in investments among members of the ruling class in the pre-industrial farming society. They preferred a life of luxury, and any money left over was saved instead of being placed into productive investments. At most, they bought more land or more slaves, but investments in the mechanization of production were seldom made. The productivity of the workers was low because of a lack of proper tools, and because they were so malnourished they could not work effectively. It was no use to raise their salaries, because workers preferred working less to earning more money. Malnourishment and disease kept the population low. War and theft regularly stopped society from building up a continuous surplus. On the whole, it seemed as if the entire society was organized to make economic growth impossible. As soon as there was a slight increase in one sector of the economy, other mechanisms soon reduced this sector to its previous level. An increase in the standard of living brought about an increase in population, which soon reduced the standard of living again. An increase in the level of education brought about a shift from productive work to unproductive activities such as administration and commerce. The efforts made to cultivate more land in order to feed the increasing population led to soil erosion and overwatered soil.

2. Why are birth-rates lower is Western Europe than in Asia? Economic historians maintain that there must be a rational explanation for this difference. Indeed, one may ask oneself why people in South-East Asia do not try to reduce birth-rates in order to reduce the number of mouths to feed.

   E.L. Jones proposes an explanation in his book *The European Miracle*. The difference has to do with the fact that the risk of natural catastrophes is much greater in South-East Asia than in Western Europe. After a catastrophe in which many people die, such as an epidemic, hurricane, or flood, people must restart their food production as soon as possible. This requires that a substantial work force is still available after the catastrophe; hence families try to produce as many children as possible so that this may be the case. This need is not present in Western Europe, where wide ranging catastrophes hardly ever occur. For example, pandemics such as the Black Death are quite rare. This pattern has been present for a long time, and the modern possibility of transporting food over long distances where it is needed has not yet affected reproductive trends.

3. In a game theoretic study, *The Peter Principle Revisited: A Computational Study* made by Alessandro Pluchino, Andrea Rapisarda and Cesare Garofalo, the authors give an explanation why bosses so often are clearly incompetent for their jobs.

   Here is the abstract, adapted from http://arxiv.org/abs/0907.0455.

   In the late sixties the Canadian psychologist Laurence J. Peter advanced an apparently paradoxical principle, named since then after him, which can be summarized as follows: 'Every new member in a hierarchical organization climbs the hierarchy until he/she reaches his/her level of maximum incompetence'. Despite its apparent unreasonableness, such a principle would realistically act in any organization where the mechanism of promotion rewards the best members and where the mechanism at their new level in the hierarchical structure does not depend on the competence they had at the previous level, usually because the tasks of the levels are very different to each other. Here we show, by means of agent based simulations, that if the latter two features actually hold in a given model of an organization with a hierarchical structure, then not only is the Peter principle unavoidable, but also it yields in turn a significant reduction of the global efficiency of the organization. Within a game theory-like approach, we explore different promotion strategies and we find, counterintuitively, that in order to avoid such an effect the best ways for improving the efficiency of a given organization are either to promote each time an agent at random or to promote randomly the best and the worst members in terms of competence.

Further Reading

Box-Steffensmeier, J. M., Brady, H. E., & Collier, D. (Eds.). (2008). *The Oxford handbook of political methodology*. Oxford: Oxford University Press.

Fay, B. (1996). *Contemporary philosophy of social science*. Oxford: Blackwell.

Martin, M., & McIntyre, L. C. (Eds.). (1994). *Readings in the philosophy of social sciences*. Cambridge: Cambridge University Press.

Ricoeur, P. (1981). In P. Ricoeur (Ed.), *Hermeneutics and the human sciences: Essays on language, action and interpretation* (trans and introd: Thompson, J. B.). Cambridge: Cambridge University Press.

Chapter 10
Scientific Laws

Nature and Nature's laws lay hid in night
God said 'Let Newton be!' and there was light

A Pope

10.1 Introduction

The goal of the sciences is to discover scientific laws, many say and according to Hempel's model of explanation we need laws for scientific explanations. Some examples of such laws are Kepler's three laws of planetary motion and Newton's law of gravitation.

Within the natural sciences it is largely uncontroversial to claim that the goal is to discover the laws of nature. Thus one might similarly think that the goal of the social sciences is to discover the laws of society. However, many have claimed, for various reasons, that there are no social laws to be discovered. Presently, I will not take a stance in regards to this issue. Instead, I shall focus on the discussion of the concept of a scientific law, taking its use in the natural sciences, in particular physics, as my starting point. As we will see, there are many deep controversies about the concept of scientific (or natural) law and e.g. van Fraassen claim that there are no laws, neither in the social nor in natural science.

But the term 'law of nature' and its associates are often used both by scientists and laymen and it is an important task for philosophy of science is to give an analysis of this concept. The questions to be answered include: What is a law? What conditions must be met for something to be called a law? Why do we sort out *some* parts of scientific theories and attribute to these a special status?

It has proven astonishingly difficult to give generally acceptable answers to these questions. John Earman has characterized the debate as follows:

> It is hard to imagine how there could be more disagreement about the fundamentals of the concept of law of nature – or any other concept so basic to the philosophy of science – than currently exists. A cursory survey of the recent literature reveal the following oppositions (among others): there are no laws of nature vs. there are/must be laws; laws express relations between universals vs. laws do not express such relations; laws are not/cannot be Humean supervenient vs. laws are/must be Humean supervenient; law do not/cannot contain ceteris paribus clauses vs. laws do/must contain ceteris paribus clauses.

L.-G. Johansson, *Philosophy of Science for Scientists*, Springer Undergraduate Texts in Philosophy, DOI 10.1007/978-3-319-26551-3_10

> One might shrug of this situation with the remark that in philosophy disagreement is par for the course. But the correct characterisation of this situation seems to me to be 'disarray' rather than 'disagreement'. Moreover, much of the philosophical discussion of laws seems disconnected from the practice and substance of science; scientists overhearing typical philosophical debates about laws would take away the impression of scholasticism – and they would be right! (Earman 2002)

In particular I agree with Earman's last remarks, that the debate is disconnected from the practice of science and that it has an air of scholasticism. I will try to avoid these defects in what follows.

10.2 Empirical Generalizations: Fundamental Laws

In the discussion about laws one may first distinguish between *empirical generalizations* and *fundamental laws*. An example of an empirical generalization could be

All grass is green, if it is alive.

An example of a fundamental law is

Energy is constant in any closed system (Principle of Energy Conservation).

Empirical generalizations are direct generalizations of singular, observable facts of the form 'the grass on my lawn is green', 'the grass on my neighbour's lawn is green', etc. Once we have collected a sufficient number of instances, and no counter-instances, we are inclined to infer empirical generality and the result is sometimes called a law. Conversely, a single counter-instance forces us to reject the empirical generalization.

Fundamental laws, on the other hand, are not bound in the same way to one type of observed phenomena. The energy principle, to take the most obvious example, is so generally applicable that it holds for the most disparate systems. Whether studying a cell, an atom, an aquarium or even our own solar system, the principle of energy conservation is valid for all. Furthermore, energy is an abstract property that cannot be measured directly, although it manifests itself in many different forms such as potential energy, kinetic energy, mass, electricity, etc. Therefore, if we were to observe an event in conflict with this principle, there are several possibilities for saving the principle from refutation. For example, one could claim that the measuring device was faulty, or that the system was in fact not closed. Summarizing, one could say that even if the distinction is not entirely clear, it is intuitive that there is a significant difference between empirical generalizations and fundamental laws.

A large number of empirical generalizations have proven to be logical consequences of more fundamental laws. A good example is the ideal gas law

$$pv = nRT, \tag{10.1}$$

which states that for any quantity of gas, the product of its pressure (p) and volume (V) is proportional to the product of its number of moles (n) and temperature (T) (R is the general gas constant.). It is a well-known fact that this law is a consequence of the energy principle, given certain simplifying assumptions concerning gas particles. But the general law of gases was discovered before the molecular theory of gases was in place; it's status was first that of an observed empirical regularity, and much later was it discovered to be a consequence of energy conservation.

Sometimes we make empirical generalizations that later prove to be false. Many would perhaps be willing to accept the proposition 'All adult swans are white' as an empirical generalization. Yet, this proposition is false, since there exists black adult swans. Thus 'All adult swans are white' is not a law.

An interesting question here is whether all true empirical generalizations are logical consequences of some fundamental laws. The answer is unknown.

10.3 Deterministic and Statistical Laws

Some laws are deterministic, and some are not. Coulomb's Law – which says that the electric force between two charged bodies is proportional to the product of the two charges and inversely proportional to the square of the distance between them– stands as a good example of a deterministic law. If we know the state of a system consisting of two charges (and nothing else) at one point of time, we can determine its state at any other time, using Coloumb's law, provided the system is not influenced by other charges. An example of a statistical law is that which concerns the radioactive decay of the Carbon-14 isotope. The probability of decay of a single C-14 nucleus during a time interval of 5730 years is 50 %. Even if we know the precise state of such a nucleus at some point in time we cannot determine its state at other times with certainty; at any point of time it may have decayed or it may not.

However, if we consider a sufficiently large number of such nuclei, more than 10^{20} or so, then we can treat this decay as the following deterministic law: after 5730 years, 50 % of them will remain. Hence we do not have to give a probability distribution given the degree of precision in such cases. However, in the case of individual atoms, we cannot know which atom will decay, or when. All we can say is that the probability that a given atom will decay within 5730 years is 50 %. Thus at the microscopic level, we can only describe radioactive decay using a statistical law.

It may seem reasonable to assume that if we had a better theory, we would be able to predict when, and under which conditions, the decay of an individual nucleus occurs. But why does it seem plausible that our present theory is incomplete? Is it not just as plausible that nature is inherently probabilistic?

Among present day atomic and nuclear physicists, the general opinion is that there is genuine randomness in nature and that the statistical character of radioactive decay does not depend on our having failed to describe all of the relevant factors. On the contrary, many believe quantum mechanics, which is the theoretical foundation for radioactive decay, to be a complete theory in this sense. However, some people are seriously convinced that nature must be deterministic. For example, Einstein is reported as once saying 'Der liebe Gott würfelt nicht' ('The good Lord does not play with dice') (One should not infer that Einstein was religious in any usual sense; it is quite clear from his writings that for Einstein this was a metaphorical way of saying that there is no randomness in nature.).

Further examples of statistical laws are some heredity laws. For example, we know that having brown eyes is a dominant trait, whereas having blue eyes is recessive. Thus if a child has inherited the gene for brown eyes from one parent, then that child will also have brown eyes. Conversely, for a child to get blue eyes it must inherit genes for blue eyes from both of the parents. If one of the child's parents, e.g. the father, has brown eyes, but one of the child's grandparents on the father's side has blue eyes, one can conclude that the father only has one gene for brown eyes. If the other parent of this child has blue eyes, then the child has a 50 % chance of getting brown eyes. This can be formulated into a heredity law in the following way:

For all x, if x is human, the probability for x having blue eyes, conditional on x having one blue-eyed and one brown-eyed parent, and one of the brown-eyed parent's parents being blue-eyed, is 50 %, i.e.,

$$\forall x P(Bx|Ax) = 50\% \tag{10.2}$$

where Bx stands for 'x has brown eyes' and Ax stands for 'one of x's parents is brown-eyed, the other is blue-eyed, and one of x's grandparents is blue-eyed'.

10.4 The Extension of the Concept of a Natural Law

By the term 'natural law' we usually intend not only things so called, such as Newton's laws, but various principles, postulates, and fundamental equations, such as the principle of energy conservation, Einstein's postulate that the velocity of light is a constant upper limit for all velocities and Maxwelll's equations. Why are these superficially quite different things grouped under the term 'natural law'?

Furthermore, there are laws, which appear to be no more than definitions of quantities. One such example is Ohm's law, U = RI, which states that the voltage (U) across a resistor is equal to the product of the resistance R and the electric current (I). Looking at the quantities that make up the SI-system–the international standard system of quantities and units–we can easily convince ourselves that Ohm's law is used to define one quantity in terms of the other two. For a long time, resistance was defined in terms of voltage and current. However, some years ago one reversed the order; with the help of the quantized Hall effect one was able to give a very accurate and operational determination of resistance. Thereafter, Ohm's law became a definition of current in terms of resistance and voltage.

The philosophical debate about laws has not addressed the question whether a definition could be a law, so I will postpone this question aside for the time being and return to this issue in Sect. 10.7. For now, lets focussing on some uncontroversial examples of laws, since it is an astonishing fact that there is deep disagreement about the concept of natural law, but common agreement about quite a number of concrete examples, such as Newton's laws, the principle of energy conservation and Maxwell's equations. But if so, do these quite different kinds of propositions really have anything in common deserving philosophical interest?

10.5 Laws and Accidental Generalisations

One might assume that, since all parties in the debate agree on a number of prime examples of natural laws, they agree on the criteria. And in fact they do to *some* extent. A formal property of all laws is generally accepted viz., they are universally generalized conditionals. This applies even to such laws as the ideal gas law or Coulomb's law, i.e. laws usually expressed as equations. The latter is given by the equation:

$$F = k\frac{q_1 q_2}{r^2}. \tag{10.3}$$

This formula is to be interpreted as 'for all pairs of bodies with charges q_1 and q_2 respectively, there is a force F proportional to their product, and inversely proportional to the square of the distance r between them'. It is implicit that this law holds only if no other forces act upon the charged bodies. Likewise, it is implicit that this law generally applies. If we were to make these implicit conditions explicit, then the complete interpretation of the above formula would be the following:

> For all pairs of bodies with charges q_1 and q_2, with a distance r between them, the force between them is
>
> $$F = k\frac{q_1 q_2}{r^2}, \tag{10.4}$$
>
> given that no other forces act upon them.

Examining the statement above, we see that (1) it is a general statement, (2) that it is of the form 'if p, then q', and (3) that it contains a version of the so-called *ceteris paribus* clause, viz., 'given that no other forces act upon them' and (4) it is true. Ceteris paribus conditions are always contained in the application of scientific laws to concrete situations, and are often implicit. However, everyone agrees that (1) and (2) are necessary conditions for all scientific laws, whereas ceteris paribus conditions are not considered constitutive of the concept of a law, even though they are often implicitly assumed.

Have we now given the necessary and sufficient formal conditions for a proposition to count as a law? No! There are many statements that fulfil these requirements and yet are not laws. One example is

All the coins in my wallet are made of copper.

(Let's suppose it is true.) This is a general statement that can easily be reformulated into 'for all x, if x is a coin in my pocket, then it is made of copper'. Thus it fulfils criteria (1) and (2); yet it is obvious that this statement is not a scientific law, even though it is true. Such statements are called *accidental generalizations.*

How does one distinguish between laws and accidental generalizations? This is the core question concerning the concept of a law, and a number of ideas for solving this problem have been proposed. We shall discuss three of those ideas here.

(i) A true law supports a so-called *counterfactual conditional*, whereas an accidental generalization does not. A counterfactual conditional is a sentence of the form 'if x, then y' in which both the antecedent and the consequent are false. Often they are expressed in subjunctive mode: 'if x were the case, then y would be the case.' For example, the sentence 'if Bill Clinton had told the truth when under oath, then he would not have been impeached' is a counterfactual, since the antecedent is false. Clinton did lie under oath about his relations with Monica Lewinsky, and he was impeached.[1]

Now compare the following two statements

1. All spheres made of gold are less than 1 km in diameter.
2. All spheres made of U-235 are less than 1 km in diameter.

U-235 is radioactive and fissible; its cricital mass is 52 kg (a sphere with diameter of 17 cm) which means that a sphere with a diameter bigger than 17 cm will start a chain reaction and explode. (This is the principle for constructing an atomic bomb.) So it is impossible to have a solid sphere of U-235 of any size. Hence (2) *must* be true and we consider it a law, since it can be derived from basic principles of nuclear physics that it is impossible to collect a sufficiently large amount of U-235 without it exploding. In contrast, (1) is not a law, even if it is true, but an accidental generalisation. It just so happens that gold is not that concentrated (And if we would find an enormous

[1] Clinton was US president 1993–2000. He had an affair with Monica Lewinsky, who had an internship at the White House.

solid body of gold somewhere in the universe, we could easily increase the limit and still get the contrast.).

From (1) and (2) we can formulate the following two counterfactuals:

3. If an object were a sphere of gold, then it would be less than 1 km in diameter.
4. If an object were a sphere of U-235, then it would be less than 1 km in diameter.

In comparing these two, we are inclined to say that (3) is false, whereas (4) is true. Thus we may state the following criterion for distinguishing between laws and accidental generalizations: If one reformulates a law into a counterfactual conditional, then the resulting sentence is true, whereas, if one reformulates an accidental generalization into a counterfactual conditional, then the resulting sentence is false.

But what grounds do we have for saying that (3) is false? If we interpret (3) as a material conditional (see Appendix), then it follows that since the antecedent is false, the whole sentence is true (See the discussion regarding conditionals in the Appendix). This shows that counterfactual conditionals cannot be interpreted as material conditionals. Thus the question regarding the conditions under which a counterfactual is true cannot be answered by investigating the truth-values of the antecedent and the consequent separately. There must be something in the connection between the antecedent and the consequent that is relevant to the truth of a counterfactual. But how shall we express this connection?

The general idea is to analyse counterfactuals in terms of possible worlds. A counterfactual is true if there exists a possible world in which both antecedent and consequent are true, otherwise false. So when we declare (3) false we mean that there is a possible world in which there is a sphere of gold bigger than 1 km in diameter.

But, then, one may ask, how do we distinguish possible from impossible worlds? A natural way do this is to say that possible worlds are those in which the natural laws are valid, but then we are back were we started. Other alternatives have been proposed, but, as Earman pointed out in the quoted passage, no idea has won general acceptance.

(ii) The second idea for a criterion to distinguish between laws and accidental generalizations is the observation that accidental generalizations refer, though perhaps only implicitly, to a particular location in time and/or space, whereas laws do not make such specifications. The American philosopher Nelson Goodman has criticized this idea by arguing that every law-like generalization is logically equivalent to a formulation that contains such a reference to space and time. The example Goodman gives is the law-like generalization 'all grass is green', which obviously is equivalent to 'all grass in London and everywhere else is green'. The difference between law-like and accidental generalizations, therefore, cannot be that the latter contains a spatial reference, since all law-like generalizations are equivalent to statements containing such references. One is now tempted to propose the following improvement: spatial reference in a law-like generalization is always redundant, whereas in accidental generalization it is not. Another way of formulating this criterion is to say that in accidental generalization, the spatial or temporal reference is necessary, whereas this is not the case as regards laws. However, this formulation immediately raises the question as to what one means by the concept of necessity as it is used in this context.

Regardless of whether this attempted solution bypasses Goodman's critique, there seems to be accidental generalizations that do not make any spatial or temporal references, as in the case of (1) above. Hence (redundant) spatio-temporal specification is not a characteristic by which one can distinguish laws from accidental generalizations in all cases.

(iii) A third alternative in the analysis of a law, which has been quite popular over the last two decades, is to claim that laws express *physical necessity*. It seems reasonable to say that it is necessarily true that bodies with mass gravitate toward each other, and that it is necessarily true that charged bodies act on one another with electrical forces. It is necessarily true that there cannot exist large concentrations of U-235, but it is not necessarily true that there exist large concentrations of gold. Thus a law has three significant traits: a law is general, it has the form 'if x, then y', and it is necessarily true.

How should one analyse the concept of *physical necessity*? It is obvious that 'necessity', as it is used here, means something other than logical necessity: there are no *logical* reasons for why charged bodies act on one another with electrical forces.

There are presently several ideas for analysing physical necessity, but none of them have succeeded in overcoming the scepticism of empirically minded philosophers. These philosophers affirm that a theory is either true or false, and the only possibility we have of determining which is the case is to compare the theory with our observations. An observation must be formulated into a statement, thus enabling us to compare the theory's empirical consequences with that observational statement. The most we can say is that the empirical consequences are true or false. However, a theory's empirical consequences are independent of whether or not we view an individual statement of that theory as a law or not. In other words, we cannot empirically determine whether a theory is comprised of laws or true accidental generalizations. But if we assume the empirical principle that there must be some empirical basis for distinguishing between physically necessary and accidental statements, then we must discard the concept of physical necessity as inessential metaphysics.

Metaphysically inclined philosophers disagree; they hold that laws are metaphysical relations between universals and we need such things in order to explain regularities in nature. Let us illustrate with Maxwell's first equation, which says that the divergence of the electric field is proportional to the enclosed electric charge. The metaphysical explanation of this would be something like: the two universals *electric charge* and *electric field* are such that in all possible worlds they are so related to each other that it makes every instance of Maxwell's first equation true (This is an application of the general idea that relations between universals are truth-makers for true sentences.). In other words, it impossible that Maxwell's equation could be false. This is, of course, a very rough description of one metaphysical explanation of the necessity of laws, but even more elaborate ones are, in my view, hardly convincing or explanatory. Do there really exist any universals, and why are they related so as to make our laws true? And more important, it is certainly not an analysis of what physicists intend when they say that Maxwell's first equation is a law.

10.6 van Fraassen's Alternative

Bas van Fraassen, who is arguably the leading empiricist of our time, has discussed, in detail, all of the main attempts at analysing the concept of a scientific law in terms of necessity and has concluded that they are all unsuccessful (see van Fraassen (1989)). He has subsequently proposed that we give up the concept of a scientific law as a metaphysical relic. He first argues that this concept is not needed in science, since all conclusions that are made can be motivated without assuming the existence of some special category of statements, *scientific laws*, which has some characteristic properties beyond truth and universality.

van Fraassen's second argument for this point is that laws can be replaced by certain symmetry principles. These symmetry principles are *conditions for our descriptions of nature*, and not conditions for nature itself. The following example may help clarify this point. When we describe the motion of a body, we must use a coordinate system and a clock. It is obvious that our choice of origin and directions in our coordinate system, as well as the choice of the starting point of the clock, is purely conventional; a question of what is the most practical. Hence it is also obvious that these purely conventional choices must not have any consequences as regards the content of our descriptions of nature. The statements we make about nature should be independent of such conventions. Another way of expressing this point is to say that our theories should be invariant, or symmetric, under a transformation from one conventional choice to another. These requirements of symmetry can be either global or local. The global symmetries are

- Symmetry under translations along the time axis,
- Symmetry under translations in space,
- Symmetry under rotations in space.

There are also a number of local symmetry principles, the best known of which is

- Phase transformation invariance of the electromagnetic potential.

These requirements may appear rather trivial, but it is interesting to notice that each such requirement of invariance under continuous transformations yields a conservation principle according to a famous theorem by Emmy Noether

(1882–1935). The well-known principle of energy conservation is one such law, as are the conservation laws for momentum, angular momentum and charge. Noether showed that each such conservation law could be derived from a symmetry principle. Thus we get the following:

(i) Symmetry under time translations entails energy conservation,
(ii) Symmetry under space translations entails momentum conservation,
(iii) Symmetry under rotations entails angular momentum conservation.
(iv) Symmetry under phase transformations of the electromagnetic phase entails charge conservation.

There are more symmetry transformations and laws of conservation, but these are the most central. It follows that these conservation laws are really just consequences of certain conditions for our descriptions of nature. These conditions are motivated by the requirement of objectivity; our descriptions of nature should be independent of the observer's frame of reference. This is hardly a principal restriction on how nature is constructed, but merely on how it should be described. It seems quite reasonable to require that our descriptions of nature be as objective as possible; that we purge all subjectivity from the universe, including how we construct our measuring devices and frames of reference. If so, we may say that conservation laws are necessary, in the sense that they are necessary consequences of objectivity demands. And general statements derivable from these conservation laws (using some definitions as auxiliaries) may also be said to be necessary and thus laws.

Conservation laws are fundamental laws, but all fundamental laws are not conservation laws (e.g. the law of gravity, Coulomb's law, and Schrödinger's equation). Thus we cannot claim that we have solved the problem of the nature of a scientific law through this analysis of conservation laws. Van Fraassen maintains that one cannot go any further if one wants to stick to empiricist principles. However, many philosophers disagree with van Fraassen on this point, even those which are not metaphysicians in the first place. In my view they have not been able to offer a more appealing analysis. In the next section I will propose an explanation of the nature of some laws, which I think would be acceptable to all empiricists, including van Fraassen.

10.7 A Proposal: Some Laws Are Implicit Definitions of Quantities

Many laws state relations between quantities. Some of these are explicit definitions, such as Ohms law, U = RI, (the voltage over a resistor equals its resistance times the current through it), and some are derivable form other more fundamental laws. But what about fundamental laws expressing relations between quantities? I will here suggest an analysis of why such fundamental laws have a special character in our theories and why we say that they are necessary. My example consists of two laws of classical mechanics.

The basic idea is that some laws at the same time are implicit definitions of new quantities and empirical generalisations of observed regularities. Their being empirical generalisations give them empirical content, and their being a kind of definition is the reason we say that they are necessarily true.

Classical mechanics is commonly divided into kinematics and dynamics. Kinematics describes the motion of physical bodies, usually called 'particles' in the theoretical exposition, since their inner structure is irrelevant, while dynamics is the theory about interactions between particles (We may treat the earth as a particle if we don't care about its volume or inner structure!). Galilei discovered some simple regularities in kinematics, for example that all bodies have the same acceleration when falling to the ground and that the distance travelled is proportional to the square of the elapsed time.

When describing particles' positions, velocities and accelerations and the relations between these we only need two fundamental quantities, *time* and *distance*. Velocity and acceleration are also needed, but they are not fundamental, they are defined as the first and second time derivative of distance. That we have two fundamental quantities in kinematics is a consequence of the fact that we need two measuring instruments, a meter stick and a clock, to measure and observe the values of the kinematic quantities. One also needs some geometry and arithmetic, of course, but these disciplines belong to mathematics; no measuring instruments are needed in pure mathematics.

How, then, do we proceed to dynamics, i.e., a theory about interactions between physical bodies? The actual history is illuminating. According to Rothman (1989, p. 85) it was John Wallis who took the first step in advancing a successful dynamics. In a report to Royal Society 1663 Wallis described his measurements of collisions of pendulums. Huygens and Wren then performed similar experiments and Newton used their results in his *Principia*. He described their experiments and findings in the section *Scholium* following corollary VI in the first section of the first book of *Principia*.

In modernized notation the result of these collision experiments is that there is a constant proportion, an observed regularity, between the velocity changes of two colliding bodies:

$$\frac{\Delta v_1}{\Delta v_2} = constant, \tag{10.5}$$

which can be written

$$k_1 \Delta v_1 = -k_2 \Delta v_2 \tag{10.6}$$

The minus sign is introduced so as to have both k_1 and k_2 positive. By testing with different bodies (which was done by Huygens) we will find that the constants really are constants following the bodies, i.e., they are permanent attributes of the bodies.

These constant attributes are their *masses*, and we may chose a mass prototype giving us the unit. So we have the following law:

$$m_1 \Delta v_1 = -m_2 \Delta v_2 \tag{10.7}$$

This is the law of momentum conservation, a fundamental law in physics. It is at the same time an observed regularity and an implicit definition of a new theoretical quantity, *mass*.

People have objected to this, saying that a proposition cannot at the same time be a definition and a generalisation of empirical observations. They are wrong. The fundamental point is that, in order to be able to formulate the generality that all bodies are such that their collisions satisfy Eq. (10.7), we need the concept of mass and this concept was not available earlier, nor is it defined by any other quantitative relation. It was invented precisely for the purpose of expressing the regularity that Wallis, Huygens and Wren found. Equation (10.7) function as implicit definition of mass (An explicit definition of mass in terms of other quantities is impossible, since it is one of the fundamental quantities in physics.).

Moreover, as Quine forcefully argued, the distinction between analytic and synthetic sentences cannot be upheld (see his *Two Dogmas of Empiricism.*[2]), and those who criticizes the idea that one and the same proposition can both be an implicit definition and an observed regularity presumably would motivate their criticism by saying that a definition is analytically true and an empirical generalisation is synthetically true.

Newton published *Principia* 23 years after Wallis reported his findings to Royal Society. He begins *Principia* by defining mass as 'quantity of matter', (on page one in the main text), but this definition only gives us an intuitive meaning of the word 'mass', it does not tell us how to measure it. Wallis, Huygens and Wren had long before the publication of *Principia* provided that without using the word 'mass'; they in effect introduced this quantitative predicate in physics.

If we now divide both sides of Eq. (10.7) with the collision time, we get (neglecting the difference between differentials and derivatives since this is of no relevance for the present argument):

$$m_1 a_1 = -m_2 a_2 \tag{10.8}$$

Let us further introduce the term 'force', labelled 'f', as shorthand for the product of mass and acceleration. This gives us Newton's second and third laws:

$$\mathrm{N2: f = ma} \tag{10.9}$$

[2] In his (1953), pp. 20–46. See also his (2004).

$$\text{N3} : \mathrm{f}_1 = -\mathrm{f}_2 \tag{10.10}$$

Thus we have got Newton's second and third laws based on an observed regularity, viz., momentum conservation during collisions between bodies.

Many readers would, I guess, oppose my saying that Newton's second law is an explicit definition of force, since forces usually are thought of as causes, viz., the causes of bodies' accelerations, which conflicts with saying that the term 'force' simply is shorthand for 'mass times acceleration'. But if so, what is here cause and what is effect? Newton's third law tells us that we can either focus attention on the motion of body 1 and ask about the force on it from body 2, or the converse), no cause-effect distinction can be made. This is also the modern view in physics, instead of talking about causes, one talks about interactions. So I reject the common interpretation of Newton's second law as a causal law; this interpretation is in fact inconsistent with Newton's third law.

Newton himself appears to have thought of forces as causes. But that cannot be correct and perhaps a remnant from Aristotelian thinking about motion.

It is thus a bit misleading to say that Newton *discovered* his second law; it is more correct to say that he changed the meaning of the word 'force', which was used in ordinary language long before Newton, when he established his second law.

What he discovered was that, contrary to the Aristotelian view that matter and quantity are distinct ontological categories, to every body he could attribute a *quantity of matter*, i.e. mass. We may carefully observe that mass is not a directly observable quantity and it is not the same as weight! This is the core idea, which entailed a radical change of our entire thinking about motion.

In this example we have talked about mechanical forces. This is no real restriction; it is well known that there is, as far as we know, four fundamental kinds of interactions in nature, gravitation, electromagnetism, the weak and the strong nuclear force, and the same reasoning applies in all these kinds of interactions. Mechanical interactions, by the way, are electromagnetic interactions.

It seems to me correct to say about science that inventing and defining new and precise concepts and establishing laws are two sides of the same coin. This is very clearly the case in classical mechanics. Neither the concept of mass, nor the modern concept of force, was used or known before Galilei and Newton; The introduction of these concepts and the establishment of the law of momentum conservation and of Newton's laws cannot be separated either historically or conceptually.

Let us now proceed to the law of gravitation:

$$F = G\frac{m_1 m_2}{r^2} \tag{10.11}$$

Since (i) we earlier have introduced the concept of mass in the law of momentum conservation, (ii) force is shorthand for mass times acceleration and (iii) distance is a kinematical quantity, it appears that this must be a purely empirical law; all concepts are earlier defined. But that is really astonishing: how could it be that four quantities, defined by other relations, always and exactly relate to each other according to Eq. 10.11? It seems to be a truly cosmic coincidence!

I don't believe in cosmic coincidences, and in fact neither do we have any such here. As Newton himself realised, the mass concept utilized in this law express another property than the mass concept occurring in our descriptions of collision experiments. The law of gravitation describes how physical bodies interact at distance, whereas the law of momentum conservation describes how bodies interact when they collide. We have two distinct mass concepts, *gravitational mass* and *inertial mass*. And we may say, just as with the law of momentum conservation, that the law of gravitation at the same time is an implicit definition of gravitational mass and an empirical generalisation found by observing e.g., how planets move around the sun and Jupiter's moons move around Jupiter.

The remarkable thing is that inertial and gravitational mass, although conceptually distinct properties, always are exactly equal! This is truly astonishing! Newton realised that gravitational and inertial mass was different properties, but had no explanation of their equality. This conundrum was not explained until the advent of general theory of relativity; and the explanation is the most natural one, viz., that gravitational and inertial mass really *is the same property*, because gravitation and inertia at bottom is the same phenomenon.

Summarising my account of classical mechanics: we have two fundamental laws, momentum conservation and the law of gravitation, which each have a double function; they are each an implicit definition of a new quantitative predicate and at the same time expresses an inductive generalisation of observations. Newton's second law is an explicit definition of force and Newton's third law is a direct consequence of momentum conservation, given the definition of force.

What, then, about the necessity of laws? First, definitions, both implicit of explicit, are *necessary conditions* for the use of the defined terms in our theories. Second, logical consequences of definitions are necessary consequences of these definitions, since necessity distributes over logical consequence. So we have an account of why we think of laws as necessary.

Medieval philosophers distinguished between necessity *de dicto* and necessity *de re*; necessity de dicto is a necessity in what is said, whereas necessity *de re* is necessity in the object, viz., that the object talked about has a property, or stands in a relation, by necessity. Applying this distinction, we see that the necessity of both the fundamental laws and of those laws derived from them, are species of necessity *de dicto*. It is the *statements* that are said to be necessary. It does not follow that the objects talked about have any properties by necessity and independently of how we characterize them. So empiricist scruples about metaphysical necessity are satisfied.[3]

---

[3] If we formalise a sentence claiming the necessity of a law as 'N "L"', i.e., treating 'necessary' as an abbreviation for 'necessarily true', i.e. a second order predicate and the sentence 'L' as an argument) one blocks the distribution of necessity into the law sentence since one cannot distribute inside quotation marks, and hence the attribution of necessary properties to things.) Cf. Quine: 'Three grades of modal involvement', pp. 158–176 in his Quine (1976).

There are, for sure, several laws that are not definitions of quantitative concepts, or consequences thereof, and this account obviously doesn't apply to these. But some progress is made, or so I hope.

10.8 Summary

Scientists and others often use phrases such as 'natural laws', physical postulates', 'fundamental principles' 'NN's law', 'NN's equation' etc. The things denoted by these phrases are often referred to as 'laws of nature' in the philosophical debate.

Philosophers disagree sharply about the concept of natural law. Some say that it is obvious that there is a certain category of propositions, *laws*, which differ from other true propositions of the same logical form, *accidental generalisations*. Many of these philosophers think that the property of being a law is based on a kind of metaphysical necessity. Other philosophers of a more empiricist bent disagree and point out that the predictive power of a theory does not depend on whether some of its propositions are called laws or not; calling some true propositions 'laws' is unnecessary metaphysics. This empiricist argument is convincing, but why, then, is the concept of natural law, used not only by philosophers but also by laymen and scientists? Surely, this concept serves *some* purpose, although perhaps not a philosophical one.

For four types of laws it is possible to give an explanation of our propensity to call them 'laws' and treat them as having a special status of being *physically necessary*, without assuming any kind of metaphysical necessity. The first type comprises conservation laws, each of which can be shown to be a consequence of a certain objectivity demand on our descriptions of the events in nature. The second type comprises some equations, each of which may be viewed as an implicit definition and at the same time an empirical generalisation of observations. The third type consists of explicit definitions of one of the quantities used in the respective equation. Finally, those true generalised conditionals that can be derived from the three types above are also labelled laws.

Further Reading

Armstrong, D. (1983). *What is a law of nature*. Cambridge: Cambridge University Press.
Bird, A. (2007). *Nature's metaphysics, laws and properties*. Oxford: Oxford University Press.
Carroll, J. (1994). *Laws of nature*. Cambridge: Cambridge University Press.
Cartwright, N. (1983). *How the laws of physics lie*. Oxford: Clarendon.
Lange, M. (2009). *Laws and lawmakers*. Oxford: Oxford University Press.
Mumford, S. (2004). *Laws of nature*. London: Routledge.
van Fraassen, B. (1989). *Laws and symmetry*. Oxford: Clarendon.

Chapter 11
Theories, Models and Reality

Structure is what matters to a theory, not the choice of objects.

W. V. O. Quine

11.1 Introduction

We use the concept of a theory essentially in two distinct ways. In certain situations, we make a distinction between theoretical and empirical propositions, where we use the concept of theory to mark a contrast with observations and things inferred from these. In other situations, the point of the concept of a theory is primarily to contrast theory with reality, observed or otherwise. It is in this second sense that we say of a hypothesis that it is 'just a theory' where we are allowing that the world may be such that the hypothesis is false. If one could claim that the content of observations are 'pure reality', then there would be no difference between these two ways of using the concept of a theory; but unfortunately, as we saw in Chap. 4, observation sentences are often theoretically loaded. One can say that when we use the concept of a theory in the first way we want to stress the abstract nature of a theory, whereas in the second usage we want to stress the provisional nature of a theory as a picture of reality.

In many cases, the concept of model is used as a synonym for theory, where 'theory' is interpreted in the second sense. In this chapter, we shall dig a little deeper into the meaning of theories as models of reality.

That a model and what it models resemble one another is easily understood in the case of a concrete object such as a toy car, but what does it mean for a theoretical model to resemble the part of reality it models? An example is helpful.

Example: Maps and Reality

The figure below shows a map of Stockholm's commuter train network. The map exhibits an essential property of the commuter train network, namely, the topology of its various connections between stations. From the map one can, for example, see that one can travel directly from Jakobsberg to Stockholm Central Station and that on the way one will pass four stations (Barkaby, Spånga, Sundbyberg and Karlberg) (Fig. 11.1).

L.-G. Johansson, *Philosophy of Science for Scientists*, Springer Undergraduate Texts in Philosophy, DOI 10.1007/978-3-319-26551-3_11

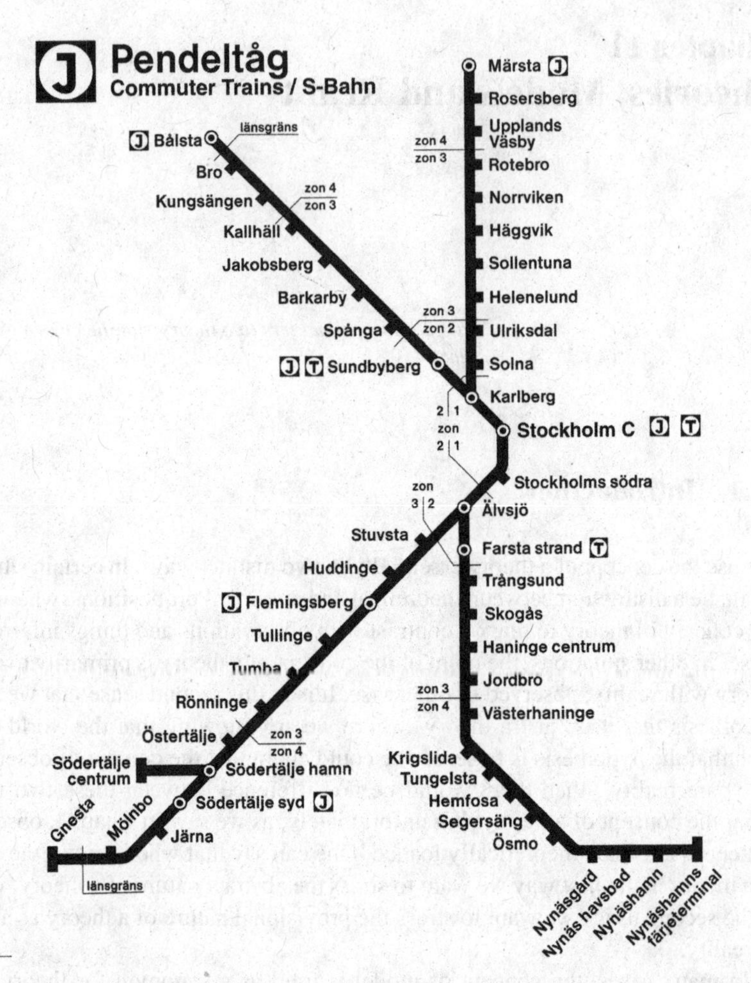

Fig. 11.1 A map of Stockholm's commuter train network

However, one cannot see how long the distance is, how windy the track is, or how long it will take to travel; the metrical properties of the commuter train network are not displayed, only the topology.

Two interesting questions immediately arise: (1) is it, in principle, possible to construct a model that gives a complete picture of a part of reality, and (2) if we had such a model, and if we knew that it was complete, should we then call it a model of reality? We shall return to these questions later on. First, we must more precisely understand what is meant by a model *resembling* reality.

I propose that resemblance must be some kind of *structural similarity* between the model and reality. Structural similarity can, in turn, be understood in two ways. In certain cases are we interested in special relations between parts of reality (things, properties, measurement results, etc.), while in other cases we are interested in certain kinds of operations or actions that can be performed on the various elements of reality. In both cases our task is to find a representation of these aspects using a model. The first type of structural similarity is the simpler of the two, so we shall begin with it. We shall then discuss the second type, which may be understood with help of the mathematical concept of *isomorphism*.

11.2 Structural Similarity as a Mapping of Relations

In some scientific disciplines theories are not mathematically formulated for various reasons. Nevertheless, one often uses models for describing reality. A common example is the use of causal models. These often take the form of diagrams where one indicates causal relations between different factors using a series of arrows. The following diagram is such a causal model, displaying the causal network behind myocardial infarction (Note that this is only an illustration and is not meant to be a correct model of the causal network!).

The fundamental structural relation in this diagram is an arrow starting at one box and ending at another. This arrow represents a causal relation; the factor in the box where the arrow starts is a cause of the factor in the box at the arrow's point.

One important feature of causal relations that is visible in this model is transitivity: if A is a cause of B, and B a cause of C, then A is an (indirect) cause of C. Following the arrows in the above diagram yields precisely this. However, what is not shown is the relative strength of the different causal factors. This is a general characteristic of all models; they display *certain aspects* of reality while leaving out others (Fig. 11.2).

An interesting question with regards to the present example is whether the model is complete in the sense that it contains all the causal factors. Generally, this is not the case as typically such models only include those causes that are known/suspected and significant. Naturally, this question is not one that can be ascertained from the model alone; rather, one must compare the model with reality.

11.3 Mathematical Models

The two models above display non-quantitative relationships between parts of reality. In contrast, some sciences make extensive use of mathematical models that give information about quantitative relationships between observed properties of objects. One example that is often discussed is the ideal gas law: the product of the pressure (p) of a gas and its volume (V) is equal to the product of the amount of matter (n), the gas constant R, and the temperature (T); $pV = nRT$. What is essential is that the model is structurally similar to (observed) reality in the form of *measurements on portions of gases*. The mathematical formula expresses relations

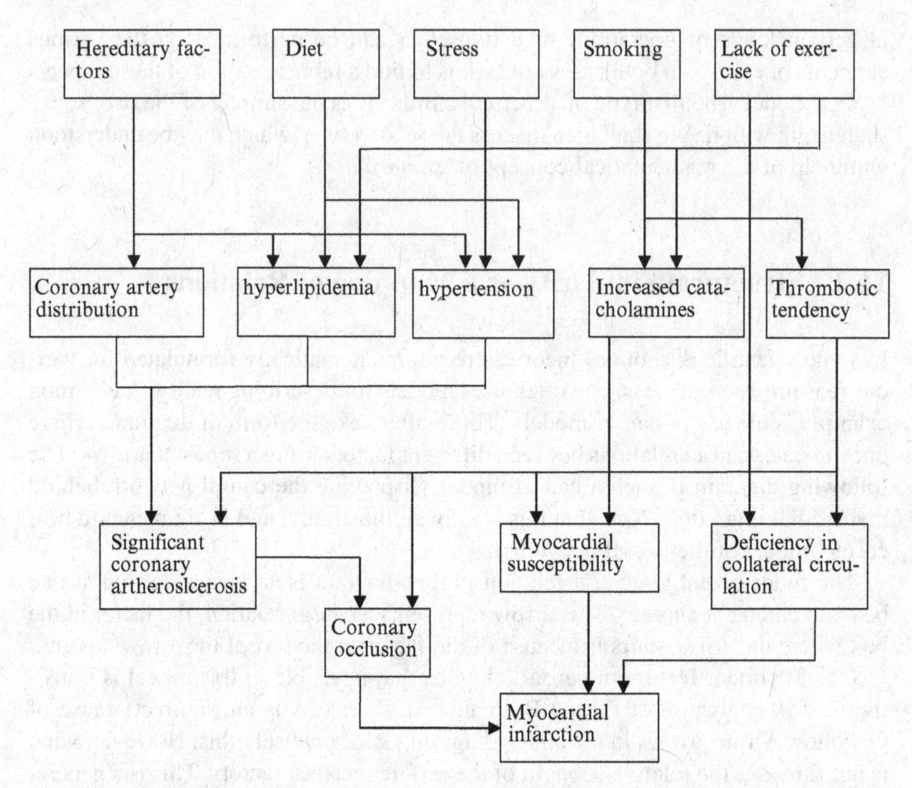

Fig. 11.2 Model of the causal network behind myocardial infarct (Adapted from Friedman, M & Rosenman, R. H. (1974))

between certain properties of any specific quantity of gas. However, it is not stated that this description of gases is complete, or even exactly correct.

The concept of structural similarity is usually specified within mathematics as an *isomorphism*, which is the idea of two sets having the same structure regardless of the nature of the elements in those sets. It is tempting to say that the observed measurements are isomorphic to the ideal gas law, but it requires a rather detailed discussion about set theory, groups, and measurements to show that. Suppes (2002) contains a detailed analysis of the notion of an isomorphism and results of measurements.

Normally, an equation functions as a model only under certain conditions. In the case of the ideal gas law, it is derived from the assumption that gas particles have no inner structure, that there are no forces active between them and that all collisions between them are elastic. However, from our background knowledge about molecules it is easy to understand that if the pressure or temperature is sufficiently high, then the gas particles will spend more time near one another. But if the gas particles spend more of their time in close proximity, the forces between them that we previously discounted will be stronger. In effect, what we have is no longer an ideal gas. Could the model be improved by removing one or more of the restrictions? Yes, in this case, taking into account interactions between gas particles gives us a more complicated law called van der Waal's law, which is valid in more cases.

Is this process of successive improvement by adding structure always possible? Most scientists seem to assume this to be the case. In fact, this assumption has been confirmed on a number of occasions, such as the addition of quantum and relativistic effects in the case of the ideal gas law. But is there a limit beyond which no more improvement of a model is possible? An obvious limit is the case where one can construct a complete model in which all observed and unobserved phenomena are captured by the model. Naturally, one can never be certain that such is the case, but if we were to have such a model, we would have reached the limit of improvement.

However, if it is not the case that our model is complete, is it then always possible to make such a model better? James Brown has presented an interesting argument for this possibility:

> We already have set theory in our possession. In some important and relevant sense we can grasp it. When we do science we in effect assert that some part of the physical world (or even the whole universe) has the same structure as some mathematical object. Since the realm of sets provides all possible mathematical structures, any way that the world could be is exactly isomorphic to some set-theoretic object. Since all of these mathematical structures are graspable in some relevant sense by the human mind, any way that the physical world could be is also graspable by the human mind. Of course, any alleged isomorphism between the physical world and some set-theoretical structure is a conjecture, which may be false; science after all, is very difficult and very fallible. But there is no way of thinking about physical reality which is ruled out by our genetically cognitive capacity, since (standard) set theory can provide the representation of any possible way reality might be. The mere fact the we possess set theory shows that there can be no (non-logical) constraints on our thinking. Brown (1994, p. 76)

The crucial point is that set theory is so flexible and general that any thing whatsoever can be represented by a mathematical object constructed using set theory. This is not the place to scrutinize the argument for this statement, let me simply state that I agree with Brown on this point. A thorough analysis of these matters are found in Chaps. 2 and 3 in Suppes (2002).

Another thing to notice is that Brown clearly hold that the world is in some definite way independent of our ideas, concepts or theories. This is the basic idea in realism. Some philosophers deny that such realist theses are meaningful or comprehendible, thus expressing a form of anti-realism. Anti-realists have further claimed that wave-particle dualism in atomic physics supports this view.

11.4 Wave-Particle Dualism

The well known phenomenon that reality's smallest pieces, neutrons, protons, electrons, photons, neutrinos, etc., show both particle and wave behaviour raises a central question; is it in principle possible to find only one model for a certain part of reality? All observable features of the micro world are captured, as far as we know, by quantum mechanics, but we need two models for thinking about the micro objects, viz., the particle model and the wave model. Thus quantum mechanics represents, in a certain sense, the dualistic character of elementary particles. This provides a strong reason to doubt the idea that theories, conceived as models, can be improved until they give a complete description of the external world. This in turn raises the following question regarding a theory's relation to reality: Are theories merely models that describe observable phenomena as the anti-realists claim, or do (the best) theories describe *reality*, not only observable traits, as the realists claim? This conflict does not concern what we know, but how the world actually is and what is meaningful to say of it.

A realist might formulate wave-particle dualism as: reality, as it were, show two faces in our experiments, the wave face and the particle face. This does not mean that the two models do not concern reality. One could still claim that we have yet to construct a more complete model, which encapsulate both wave and particle behaviour in one model.

However, according to Bohr, this is not possible; nature has *complementary* properties, which makes the completion of either model so as to cover all phenomena, impossible. That two properties are complementary means that the *conditions* for observing these two properties in the same system are mutually exclusive. It is important to note that the properties themselves are not contradictory, but the *conditions for ascribing these two properties to a system* are. Concretely, this means that in certain situations, when we arrange for observing positions of particles, which are decided by how we arrange our experiments, the conditions for observing wave behaviour are not fulfilled and cannot be fulfilled. In other situations, where the conditions for observing particle behaviour are not fulfilled but the conditions for wave behaviour are, we observe only wave behaviour. It is not possible to conceive of situations arranged such that both types of properties are observable, according to Bohr. The conditions for observing particle behaviour and wave behaviour are logically incompatible.

One is now prone to ask whether the objects under observation do not simultaneously *have* both wave and particle properties (which would explain why we cannot observe them at the same time), or whether it is only a principal limit of our ability to measure and observe the micro cosmos that hinders us from simultaneously observing particle and wave behaviour. This is a controversial question in the philosophy of physics. I believe that the first alternative is the correct one. In short, in motion from one place to another all electrons, protons, photons, etc., have wave properties (wavelength, frequency, extension in space), while during interactions (collisions) they have particle properties (definite position, energy, momentum). This makes it possible to construct a model that fulfils the requirement of structural similarity to the reality that presents itself in all of our observations. And this has been done! That is what we have in quantum mechanics. This theory is a *mathematical model*, which, as far as we know, represents all observable phenomena in the micro domain. Antirealists have always claimed that this purely mathematical structure cannot be given one unified visualizable interpretation in terms of things moving around in space, hence, we should accept that realism is disproved. This is a highly controversial issue. I do think quantum mechanics, as we now know it, can be given a visualizable interpretation, but this issue is far beyond the scope of this chapter.[1]

11.5 Can One Measure Structural Similarity?

In some areas of research and technology, such as image reproduction, one is interested in measuring how good a copy is or how structurally similar two different objects are. Could it be measured? In the case of similarity between images we have the SSIM index, which measures the structural similarity between two images. It is valued between -1 and 1. When two images are nearly identical, their SSIM is close to 1, see Wang et al. (2004). The function *ssimval = ssim(A, ref)* in Mathlab computes the SSIM value for a picture A, with 'ref' as reference picture.

Another measure of structural similarity, to be applied to graphs displaying business processes, is presented in Dijkman et al. (2009).

11.6 Ontology and Structural Similarity

Suppose that we have succeeded in constructing a mathematical model of reality that is perfect in the sense that all observable phenomena fit in with this model. How do we know that the model is not reality itself? The answer may seem obvious. Models and reality are entirely different types of things. In many cases, a model consists of mathematical objects that *represent* real things and their properties. This is the fundamental relation, and it is obvious that we could not talk about something representing something else if we could not distinguish the two relata. The fundamental difference is that the elements of the model are mathematical objects, whereas the elements of reality are physical objects, things that can be found in space and time.

Is this difference between model and reality actually important for how we think about models and theories? Quine has claimed that 'structure is what matters to science and not the choice of its objects' (*Theories and Things*, p. 20.). What he means is that from a scientific perspective it does not matter whether we suppose that physical reality is built of particles, fields or even some type of mathematical object (n-tuples of numbers) so long as the sets of these objects have the same structure. This stance also implies that one can assert the existence of abstract things like numbers, functions, social institutions, etc., without dwelling in speculative metaphysics. The essential step in constructing theories about, e.g., social

[1] I have given my view on quantum mechanics in my (2007).

institutions is that we clearly state how these postulated entities relate to each other and to statements about observable phenomena. However, science cannot answer the ontological question of what there is in the world, *independently of which theories we use*; we can only say things like, 'according to theory T, there exists objects of kind K, and since we have very good evidence for T and it is presently the best theory available, it is reasonable to assume that there exists K-things.' Now assume that we make progress and construct a better theory T′, which assumes not objects of kind K but other kinds of objects. So we have changed our ontology, but what we have not changed is some structural features, in one of the senses here described. So structure is what really matters. This is not an uncontroversial stance, quite the contrary, but the arguments in its favour are stronger than one might think at first glance.

Further Reading

Brown, J. (1994). *Smoke and mirrors* (How science reflects reality). London: Routledge.

Dijkman, R., Dumas, M., & Garcia-Banuelos, L. (2009a). Graph matching algorithms for business process model similarity search. In U. Dayal et al. (Eds.), *Business process management*. Berlin: Springer.

Giere, R. (1999). *Science without laws*. Chicago: Chicago University Press.

Ladyman, J., & Ross, D. (2007). *Every thing must go: Metaphysics naturalized*. Oxford: Oxford University Press.

Quine, W. V. O. (1981). *Theories and things*. Cambridge, MA: Harvard University Press.

Suppes, P. (2002). *Representation and invariance of scientific structures*. Stanford: CSLI.

Suárez, M. (2003). Scientific representation: Against similarity and isomorphism. *International Studies in the Philosophy of Science, 17*, 225–244.

Wang, Z., Bovik, A. C., Sheikh, H. R., & Simoncelli, E. P. (2004). Image quality assessment: From error visibility to structural similarity. *IEEE Transactions on Image Processing, 13*(4), 600–612.

Part III
Some Auxiliaries

Chapter 12
The Mind-Body Problem

> *'O blithe little soul, thou, flitting away,*
> *Guest and comrade of this my clay,*
> *Whither now goest thou, to what place*
> *Bare and ghastly and without grace?*
> *Nor, as thy wont was, joke and play.'*
> *Hadrian, on his deathbed (Translated by A. O'Brien-Moore)*

12.1 Introduction

What is the relation between the body and the soul, or mind? Is the mind something that exists independently of the body? Or are the body, specifically the brain and nervous system, and the mind merely two different descriptions of the same thing? Our common sense understanding, which is riddled with religious influence, leans towards the first alternative. But then what is the mind? How do the mind and the body interact, if they are in fact fundamentally different things?

If one assumes the former position, one is a dualist and if one assumes a variant of the latter position, one is a monist. There are many versions of monism but basically only two versions of dualism, substance dualism and property dualism.

Dualists and monists face completely different requirements for explanation. If one believes that body and mind are not two different things, but rather one and the same, then what is wanted is an explanation of how mental processes occur in the brain. Such an explanation requires a reduction of the mental to the physical. There are several examples of successful reductions that illuminate how this requirement might be met in this case. One such example is the reduction of genes, hypothesized by Mendel as the carrier of inherited traits. Originally, this concept was a purely theoretical construction. After the discovery of the DNA molecule in the 1950s, one was able to guess that genes were parts of nitrogenous base sequences in the DNA molecule. This was latter confirmed when specific gene sequences were identified along with their phenotypic expressions. As this is an on-going investigation, the details of the reduction are still to be worked out; but one can justly say that the problem of reducing phenotypic traits to heritable genes, now conceived as sequences in DNA molecules, has been solved in principle.

The monists think that an analogous reduction of mental phenomena to physical processes in the brain is at least in principle possible, while others hold that

L.-G. Johansson, *Philosophy of Science for Scientists*, Springer Undergraduate Texts in Philosophy, DOI 10.1007/978-3-319-26551-3_12

reduction, in the sense of identification types of mental states with types of physiological states, is impossible. But this rejection of the possibility of reduction must not be taken as an argument for dualism. One position, eliminative materialism, holds that there is no scientific role for mental phenomena to play; hence they can be eliminated from a scientific description of the human being. The argument justifying this elimination is based on logical and semantic analyses of the reduction concept, as well as analyses of the concept of a mental state.

Nowadays, all monists are materialists; that is, people who think that external reality is material in nature (The concept 'material' should be understood here in terms of modern physics. Since energy and matter are equivalent, according to the well-known formula $E = mc^2$, even energy belongs to the material world.). The opposite thesis–that external reality is mental in nature–is also a form of monism. This latter view is very unusual in recent times, but has had some significant proponents, for instance, Leibniz and Berkeley.

Dualists face a completely different explanatory task; they must explain how it is that the body and the mind can interact. All hitherto known theories of how objects interact are strictly physical in that they describe how physical objects interact via one of the four fundamental natural forces, gravitation, electromagnetism, weak and strong nuclear interaction (Materialism is nowadays identified as the thesis that ultimately all that exists are physical in nature.). If the mind is a non-material object, then no purely physical theory about its interaction with the body can be given. Indeed, the idea that the mind is non-physical, together with the observation that bodily and mental states can causally interact with each other, conflicts with a basic tenet of our scientific thinking, namely, that the physical world is causally closed. Nothing non-physical, if such a thing exists, can influence something physical (For example, abstract things like numbers are non-material things and they cannot possibly be causally related to anything in the physical world, in fact not to anything whatsoever.). This idea is seldom expressed explicitly, probably because most take it to be obvious and without need of conscious expression; but nevertheless, it is fair to say that it is a firmly held tenet of the sciences.

12.2 Substance Dualism

Substance Dualism is the theory that there are two types of substances, or entities: bodies and minds. Property Dualism – an alternative dualist theory most notably championed by Popper[1] and David Chalmers[2] – stands in marked contrast to substance dualism in insisting that there is only one type of substance in the

[1] Karl Popper, 1902–1994, was born in Austria but worked in England beginning in 1945. His central works falls within the philosophy of science (see Sect. 12.2).

[2] Chalmers, D. (1996). *The conscious mind. In search for a fundamental theory*. Oxford: Oxford University Press.

world, matter. Property dualism deserves its name because this single substance has two irreducible types of properties: physical (including chemical and biological) and mental. I shall discuss the latter view in the next section.

Descartes proposed the most well-known substance dualistic theory.[3] He claimed that man is constituted of matter and soul, two different types of substances, each having a fundamental property, or essence. The essence of bodies is extension, and the essence of minds is thought.

Descartes' explanatory task was to show how mind and body interact. Starting from the new mechanical world-view of his time, that all motion is the result of mechanical interaction, push and pull, which occurs when portions of matter, bodies with extension, come into contact. In other words, only things with extension can put a piece of matter into motion. Yet it seems obvious that many of our own bodily motions are caused by our desires and thoughts, our mental processes in our soul, and our soul lacks extension.

Descartes' solution to this problem is to state that the soul is located in the pineal gland and that mental processes set the pineal gland into motion. The motion of this gland is forwarded to the rest of the body via the so-called 'animal spirits', a viscous fluid located in the nerves (Descartes, like all others at the time, believed that the nerves were hollow and filled with 'animal spirits'.).

But why would the soul set just this particular gland into motion? Descartes had two arguments for this. Firstly, since we have only one soul, it must be located in a central part of the brain that has no duplicate, i.e. the pineal gland. Secondly, there is a cavity surrounding the pineal gland, which explains, according to Descartes, how the pineal gland could stimulate motion in the body via the fluid, the animal spirits, that he believed filled both this cavity and the nerves.

A moment of reflection tells us that Descartes' explanation breaks with the principles he himself held: if the mind can set the pineal gland in motion, then there is an exception to the principle that all bodily motion is caused by purely mechanical interactions with other bodies. Thus his theory is self-contradictory.

There are still some people who believe in some version of substance dualism, but none has been able to explain how interactions between mind and matter work.

12.3 Property Dualism

According to Popper, there are three realms: World 1, World 2 and World 3. World 1 is the material world of bodies and physical objects, and World 2 is the matter-independent world of mental processes. World 3 contains such abstract objects as

[3] René Descartes, 1596–1650, was a French philosopher who spent his last year in Sweden working for Queen Kristina. He broke with the scholastic tradition of philosophy, using systematic doubt as a method for arriving at knowledge.

scientific theories, musical works, novels, etc., i.e., cultural products that cannot be identified with physical objects.

According to Popper, entities in worlds 2 and 3 have properties that cannot be explained or reduced to objects and properties in world 1. When a material system becomes sufficiently complicated, new mental, non-material properties arise. Popper has not given any explanation for how or why they arise in a sufficiently complicated system and, without such an explanation, this theory seems to create more problems than it solves.

David Chalmers has in his (1996) proposed a somewhat similar theory. He argues that the mental cannot be reduced to the physical; that is, human (and perhaps even animal) consciousness is a radically different kind of property than the physical properties used to describe the brain. What we need is a completely different kind of theory, one that makes room for both physical and mental properties. Chalmers does not claim to have such a theory. Rather, he claims only that all previous attempts at reduction have failed, and must fail. Thus the only way out is to postulate two fundamentally different types of properties as part of an as yet unknown true theory. The first problem that Chalmers focuses on, in regards to this endeavour, is the so-called 'problem of qualia', which we shall discuss in Sect. 12.6.

12.4 Monism

The principal difficulties of dualism have resulted in that virtually all present-day philosophers are monists as regards the mind-body problem. However, this does not mean that the philosophical debate surrounding this issue has ceased, quite the contrary; there is a great diversity of views within monism. Neither does this mean that one denies the existence of mental properties; rather, it is typically thought that mental properties are, in some sense, dependent on physical properties. Though, some do go further and deny the existence of mental properties. In any case, the least common dominator of monism can be characterized by the following three theses:

- *Supervenience:* If two objects have exactly the same physical properties, then they have exactly the same mental properties (But not vice versa.).[4]
- *Anti-Cartesianism:* Nothing can have only mental properties.
- The physical world is *causally closed*. If x is a physical event, and y is either a cause or an effect of x, then y is also a physical event.

Comments and Illustrations

The first principle may seem empty, since it is not likely that there exist two objects that have exactly the same physical properties. Yet, even if this is the case, it would

[4] There are also other characterisations of supervenience, but this will do for our purposes.

not constitute an objection, since if we reformulate the thesis to its contrapositive, we get the following: if two objects differ in mental properties, then they also differ in physical properties. Even more relevant is the application of this thesis to successive states of a single person: if a person experiences a mental change, then she has undergone a physical change. If, for example, a person were to change mental state by having a new thought, then this person must also undergo a physical change. This is what it means for mental properties to supervene upon physical properties. Supervenience also seems to presume that there are properties, but this is incorrect: those who generally deny the existence of properties, nominalists, can express the supervenience doctrine in terms of classification of states using physical and mental predicates, without assuming that predicates refer to properties.

Anti-cartesianism is the thesis that there doesn't exist any purely mental objects. It implies, among other things, that the soul cannot live after the body dies. It also conflicts with the idea of reincarnation; that a mind, or soul, with certain properties can transfer from one body to another.

At first glance, the physical world's causal closure may seem to conflict with everyday observation: is it not obvious that the physical events we observe give rise to sensations and thoughts? Do not physical events appear to have mental effects? This would be a strong argument if one presupposes that sensations, emotions and thoughts have mental properties only. But if one accepts supervenience, then this argument fails, since the act of having a sensation, emotion or thought is to undergo a change of mental state, and that is a change of physical state. Thus the question of how physical and mental changes relate to one another is a question about the details of monism, for which there are quite different views.

12.5 Monistic Theories

1. Chronologically, the first serious monistic theory was *identity theory*, first proposed by J. J. C. Smart in 1959. According to this theory, when we talk about a mental state, what we are actually talking about is a certain physiological brain state. For example, feeling happy, or perceiving the sky as blue, is nothing more than the brain being in a certain state. Thus the mind-body problem will be solved once we have matched up every mental state with its corresponding physiological brain state. A historic parallel from astronomy will help illustrate how empirical research can result in propositions of identity.

   In oldest times it was believed that the Evening star and the Morning star were two different stars, but the Babylonians discovered that they were actually one and the same planet, Venus. The expressions 'Morning Star' and 'Evening Star' thus denote the same thing. Analogously, identity theory monists hold mental and physical predicates to describe the same events.

   There are two variations of identity theory: *type identity* and *token identity*. Type identity states that for every *type* of mental state there is a neurological/ physiological *type of state* that is identical to that type of mental state. This means that every time I have the sensation 'seeing that the sky is blue' I am in the same type of brain state. Token identity claims that every token of a mental state is identical to some token of a brain state, but it is not always the case that e.g. all instances of 'seeing that the sky is blue' corresponds to exactly the same type of brain state. Thus there can be identity at the token level without there being identity at the type level. Another way of expressing this is the following: physical and mental concepts classify phenomena is different ways, though they do classify the same phenomena; namely, the neurological states of a person.

An influential theory that further develops token identity is Donald Davidson's's[5] anomalous monism, but more on this in Sect. 10.7.

2. *Functionalism.* Functionalism was once proposed by Hilary Putnam[6] and many philosophers and researchers in artificial intelligence have followed suit. According to functionalism, a mental term does not refer to any definite event or neural process; rather, it indicates a place in a functional system. Thus mental terms refer to functions in the central nervous system, functions that are carried out by processes in the brain. However, mental terms do not describe these processes in terms of which neurons are active or how they interact, but rather in terms of their input and output signals and what they do; their function. Consequently, the main difference between the functional and identity theories is that a certain function can be executed by different neurons in different ways.

   One can say that the concept of a function and the concept of a process of the brain are both used to describe a certain part of reality, that which goes on inside our heads, but that these concepts classify this part of reality in different ways. Concepts such as brain processes and brain states classify using physiological criteria, whereas functions apply functional criteria. The following example illustrates this point.

   Assume that I believe it to be cold outside. This belief makes it so that I react to a certain input in a certain way. If someone were to ask 'how is the weather?' (input), I would answer 'it is cold' (output). A different output arises if the circumstances are different. If, for some reason, I were to go on a walk, then my belief that it is cold will result in my putting on warm clothes before I go out. Saying about someone that she believes 'It is cold outside.' is a description, in mental terms, of a large set of such input–output functions realized in the body of this person.

   Functionalists do not deny that these input–output functions are performed by nerve cells in the brain, but the state we describe with the expression 'believing that it is cold outside' does not describe any physiologically defined type of brain state. It is fully possible that completely different sequences of states can bring about a connection between an input of the type 'someone asks about the

[5] Donald Davidson: *Essays on Actions and Events*. Clarendon Press, 1980, especially essays 11 and 12.

[6] Putnam (1960, 1967).

weather' and an output of the type 'putting warm clothes on when you go out'. This implies that every system (e.g. a computer) that reacts similarly to a human with respect to some stimuli is, by definition, in the same state of consciousness as a human (Hence research in AI, artificial intelligence, is heavily influenced by functionalist views on the mind.).

A number of versions of functionalism have emerged, what is given above is only a first rough description. For a thorough overview of the different positions, arguments and problems, see http://plato.stanford.edu/entries/functionalism/

3. *Eliminative materialism*. According to this view, propounded by e.g. Paul and Patricia Churchland, our mental vocabulary is misleading, because it presupposes that there exist such entities as sensations, feelings, and beliefs. In other words, use of mental concepts makes us think that they refer to well-defined objects or processes, and this need not be the case. Once we have achieved a complete science of consciousness, it will contain no such concepts, but rather other, scientifically defined concepts. That is to say, we would have eliminated our traditional psychological vocabulary. This theory can be considered a variation of token identity theory, if we add the additional assumption that the everyday mental vocabulary can be eliminated from a real theory.

   Another way of expressing this is the following: if we assume that a reduction (to be explained in the next section) of mental properties to physiological properties is, in principle, impossible, then we will have, if and when this reduction is performed, a number of identity statements of the form 'mental state description M and physiological state description F denote the same state'. These identity statements constitute a sort of psychophysical law. Eliminative materialism denies that such laws exist because the mental state description is empty.

12.6 Three Important Problems for Reductionists

Reductionism in general comprises of at least two theses: (i) concepts in the reduced theory should be defined in terms of concepts from the reduction basis, and (ii) that laws in the reduced theory can be inferred from the base theory. These conditions do not follow from the three monistic principles supervenience, anti-cartesianism and causal closure, thus allowing one to deny that reduction of the mental to the physical is possible and still be a monist. But reductionism about the mental faces severe problems, which is the reason why quite a few philosophers accept monism but are sceptical about reductionism.

Arguments against reductionism are usually based on three purported distinctive features of the mental that are held to raise profound difficulties for reduction. These features are intentionality, i.e., the property that mental acts are directed towards objects (see Sect. 5.2), subjectivity and qualia.

(i) *Intentionality*. Brentano defined the mental sphere in terms of intentionality. All mental phenomena are intentional, whereas no non-mental phenomena are intentional; intentionality is the distinctive trait of mental phenomena, according to Brentano. As explained in Sect. 5.2, a mental phenomenon is constituted by two aspects, an *act*, which is directed towards an *act-object*; intentionality is this relation of directedness from the act to the act-object. A perception of an object is directed toward the object observed. When we think about a thing, person, event or theory, this thought is directed towards these concrete or abstract objects. When we are mad at someone or something, we feel *about* that person or thing. In all of these cases, the intentionality is indicated by the preposition that points to just that at which the mental process is directed.[7] Indeed, most, if not all, mental processes have this intentional aspect, whereas no material things or processes do.

Possible exceptions to the rule are moods such as euphoria, depression and anxiety. These states are similar to certain emotions, and yet they lack the directedness of those emotions. But one could defend Brentano's stance by saying that these states are not really mental states, they are physiological states, which can be clearly identified by physiological criteria. However, for the present discussion we need not decide the matter.

But how can one reduce the mental to the material? Given that the act of reduction removes an important property of the original phenomenon, must not all forms of reduction involve loss, and so be destined for failure?

Classical reductionism is the thesis that a successful reduction consists in an analysis of the semantics of mental expressions in terms of non-mental categories. For example, functionalists think that the mental state of believing 'It will be snow tomorrow.' cannot be analysed into two separate parts – the act of believing and the content of that belief – but rather that the state description, in its totality, refers to a certain functional state. A certain mental act, such as the expectation that the weather will be nice tomorrow, is a certain functional state, but the same act with different content, e.g. the belief that it will rain tomorrow, is obviously a different mental state. There need not be anything in common, indicated by the psychological word 'believe', to these two states. The intentionality, i.e. the act-object structure, here has, so to speak, disappeared from the analysis. Whether this is a failure or a success of the analysis is debatable.

(ii) Another aspect that disappears in the process of reduction is the purported qualitative aspect of (but not limited to) experiences. When I see a red tomato, my experience of that tomato contains a special chromatic quality, which I recognize and describe as 'the experience of the colour tomato-red'. This special experiential quality seems to be missing when we describe the experience in physical terms. We can perhaps achieve a reduction of the sensation

[7] There is a profound difficulty with this theory, viz., how to analyse the relation act – object in cases of non-existing objects, such as Santa Claus. This difficulty inspired Husserl and Frege to develop their theories in quite distinct ways, which gave rise to the division between 'continental' and 'analytic' philosophy.

of a smell, taste or vision by identifying it with a chain of signals in the brain. For example, we know that the sense of taste operates with five basic tastes (sour, salty, sweet, bitter and umami) and that all tastes can be described as combinations of these components in various magnitudes. We can thus represent a certain taste using a quintuple. For example, the quintuple $<1, 4, 7, 3, 8>$ can be interpreted as '1 part sour, 4 parts salty, 7 parts sweet, 3 parts bitter and 8 parts umami'. It is even possible to identify such quintuples with signals in the part of the brain responsible for the sense of smell. However, the purely qualitative aspect of the sensation, how it *feels* to smell wild strawberries, or how it feels to see red cherries (not the associations we often attribute to the sensation, but the sensation itself), also seems to disappear in the process of reduction, regardless of which form we apply.

We should note that the qualitative aspect plays no role in the use of qualitative words, such as 'red' or 'the smell of wild strawberries'. Each child must learn her native language by interacting with competent language users and the only way to learn the correct use of for example the word 'red' is to observe the situations in which adults indicate the correct and incorrect use of that word through their behaviour. This requires that the child be able to identify the common perceptual traits – e.g., the presence of red objects – in all situations where the word is used correctly. In turn, this requires that the child's is able to compare an actual sensory impression with the memory of past sensory impressions. Thus it is necessary for learning the correct usage of certain words that we experience qualitative differences and similarities between various sensations, and that we compare past and present sensations. The child must be able to remember sensations of colour hues and being able to classify a present sensation as more like one rather than other remembered colour experiences.

However, this is obviously a process that occurs within a single person; there is nothing that guarantees that two people have the same, or even similar, experience when they see a red flower. The only requirement for successful verbal communication is that different people apply category words similarly. Whether different individual's qualitative experiences differ makes no difference as regards their classification.

Many philosophers have taken for granted that humans have similar, qualitative experiences, but such philosophers are immediately faced with the question of what is meant by 'similar' in this context. In order to meaningfully say that different person's experiences are similar, or not, we must have a similarity measure that can be applied interpersonally. But such cannot in principle be had, since we are things that by definition are irreducibly subjective.

If we were to formulate these criteria in terms of the ability to report the same thing and make similar distinctions in language, then we would have transformed the qualitative aspect into a matter of behaviour, which is, as we saw above, something different than the experience itself. But how else should we formulate a criterion determining whether two individuals have qualitatively similar subjective experiences? If one then abandons the idea of giving any identity criteria, one also abandons the idea that qualitative experiences are comparable in any way.

However, we might be able to say *something* about qualia; namely, we can typically distinguish between cases where qualia are present and cases where they are not. We can easily imagine two people eating wild strawberries, where one of which has a nasty cold. One of them says to other, 'these really taste like wild strawberries', and other responds, 'sorry, but they don't taste of anything at all'.

Is it conceivable that a robot without consciousness could behave in a fashion similar to the people in the above example? Can we conceive of a robot that could carry on a conversation with humans, behave like humans, at least in some situations, without having any sort of consciousness, or self-awareness? Assume that we have succeeded in constructing a prosthetic tongue and nose, which are wired to a computer. These artificial organs can identify tastes and smells almost as well as a human (Suppose that we know a great deal about the senses of taste and smell.). Signals are forwarded to the computer, which processes the information through subroutines, thus generating applicable statements that are sent to the speech organ for communication. Nothing in this example seems impossible; and yet there is nothing that indicates that the computer truly feels the taste of the wild strawberries. Thus in the above example we have a robot that behaves like a human without having consciousness. It is not difficult to generalize this example to the entire spectrum of human behaviour. Doing so would give us a zombie, a robot that behaves like a human but is devoid of an inner life. According to critics of reductionism, this shows that any such reduction looses the most essential aspect of conscious processes: consciousness itself.

Eliminative materialists reply that if we can give an explanation of human behaviour without needing to take into account that some part of our behaviour is intimately related to conscious sensation and thought, then this aspect is irrelevant to scientific theories about consciousness.

The functionalists, on their part, stress that if consciousness has a function, then it can be described in terms of certain input–output functions, which are active when we are conscious and not otherwise. Furthermore, it is the reductionist's goal to identify these functions. The purely qualitative aspect of conscious states, 'how it feels...', is irrelevant in both cases.

(iii) All mental states are subjective in the sense that no one other than the person experiencing that state has direct/introspective knowledge of them. Contrarily, all descriptions of brain states are public, since any observer with the requisite knowledge and training can observe them. However, this leads to problems for the identification of mental states with brain states. As regards identity theory, the problem is quite clear. If two descriptions–one in mental terms and another in neuro-physical terms–describe one and the same phenomenon, then this phenomenon must have the same properties regardless of the description, otherwise the two descriptions do not describe the same thing. However, if something is public then it is not private, and vice versa. Thus all mental phenomena have a property that all neuro-physical phenomena lack.

In my view, the most serious of the three problems for reductionism is intentionality. Mental directedness appears to be an essential and fundamental property of mental activities. Every monistic theory must therefore be able to say something about what this intentionality amounts to in physiological terms, or else convince us that the act-object distinction is merely a peculiarity of our grammar.

The subjective nature of mental activities does not seem to be as fundamental, or at least not in the same way. Suppose that the strong form of (type) identity theory were true (I do not think that this is particularly likely, but the point is to show that subjectivity is a contingent fact.). Assume further that we have succeeded in pinpointing exactly which systems in the brain are active when we have a specific sensation. One would then be able to know whether or not some person has this sensation by measuring the activity in these systems using an MRI machine for example. This would mean that our states of consciousness are no longer subjective, but rather fully accessible to physiological study.

In regards to the problem of qualia, I doubt it is a real problem for reductionists. Science aims to describe and explain phenomena in objective, i.e., intersubjective, terms. Thus giving a scientific characterization of what the purely qualitative aspect is, seems to conflict with the basic point of science. The purely qualitative content, the 'subjective feel', of the conscious experience, cannot as a matter of principle be described. Why so? Because using words, language, is essentially an intersubjective activity and the word 'qualia' is intended to refer to that which is not intersubjective, but purely subjective. But that is impossible, so the word 'quale' cannot have any reference. Hence, not even in the mental sphere is there anything which we refer to by using the term 'qualia'.

Wittgenstein's discussion of these matters in his *Philosophical Investigations* will perhaps illuminate this issue. In § 304, he discusses pain, or more precisely, the experience of pain, imagining an interlocutor that argues with him:

> 'But you will surely admit that there is a difference between pain-behaviour accompanied by pain and pain-behaviour without any pain?' – Admit it? What greater difference could there be? – 'And yet you again and again reach the conclusion that the sensation itself is a nothing.' – Not at all. It is not a something, but not a nothing either! The conclusion was only that a nothing would serve just as well as a something about which nothing could be said. We have only rejected the grammar which tries to force itself on us here.

I interpret this passage as saying that an individual pain experience cannot be described as being such and so; one cannot say anything more about the experience of pain than that one is, in fact, experiencing it. If so, it cannot be treated as an object for predication.

Suppose I have headache and tell that to someone else. Now why does Wittgenstein say that we cannot say anything about it? Why isn't e.g., 'headache' a description of a certain state? His point was, I think, that if a description is to have any use in a sentence, it must be possible state criteria telling whether an object satisfy this description or not and that these criteria must be such that people can agree whether they are satisfied or not. But if these criteria in principle cannot be intersubjectively applied, as is the case of qualitative experiences, then such a description has no role to play in our language. It would lack meaning. And one might doubt if it is legitimate to talk about an object about which we cannot say anything.

What one can do, however, is categorize with reference to external objects or circumstances. We can distinguish between experiences of red and experiences of blue because the criteria for this distinction can be formulated in terms of the colour of external objects. The same goes for pain (We can distinguish between pains in different parts of the body, for example.).

The idea that Wittgenstein wants to discard is that, since 'pain' is a noun, it must refer to a thing, an object, about which one can make true or false propositions. According to Wittgenstein, this idea is incorrect. The conclusion is thus that the words we use to talk about inner states cannot function as terms referring to objects of any kind. In Wittgenstein's words we should reject the grammatical idea that nouns must refer to objects.

Some philosophers, such as Chalmers, have objected to the above conclusion by arguing, that we need a better theory: our present understanding of mental phenomena is severely incomplete.

A better theory requires new concepts enabling us to formulate more detailed descriptions of mental states. But Wittgenstein's point remains. Concepts by their very nature require criteria for application that can be intersubjectively determined, and if we by the word 'quale' or 'the quality of a phenomenal experience' try to refer to a certain state of mind that by definition is subjective and not accessible to other persons than the person who is having the experience, then we have failed to refer to anything.

Thus, of the three purported features of the mental, intentionality, subjectivity and the qualitative aspect of conscious states, I think the one to really take care of for reductionists is intentionality. Subjectivity may not be irreducible and I think we should dismiss talk about qualia as a philosophical mistake; talking about qualia is trying to say something where nothing could be said.

12.7 Mental Causes

Light, sound waves and molecules constantly bombard our sense organs, causing sensations to occur in us, whether we are awake or asleep. Sensations are mental states that, together with memories, cause other mental states, such as emotions or expectations.[8] These emotions, along with certain beliefs, cause our actions, or so it

[8] In Brentano's distinction between physical and and mental states, sensations are physical states since they have no act-object structure, which for Brentano was the distinctive trait of mental states. But here we include sensations as pains in the mental.

is generally thought. These actions often have a physical component; we move our arms and legs or emit sound waves. According to this description, physical events cause mental states, which in turn cause other physical events. But how can this be? Recalling what was claimed in Chap. 7, if an event is *the cause* (or *a cause*) of another event, these events are connected to each other by one of the four fundamental types of interactions found in nature. Thus every causal connection between two events must be a connection between physical events; as previously discussed, the physical world is causally closed. However, a mental event is not a physical event, so it seems that we have a problem.

There are three conceivable solutions to this problem. First, one could give up the principle regarding the causal closure of the physical world. Second, one could deny that the concept of a cause is applicable to the connection between thoughts and feelings and our actions, and third, one could say that mental events are in fact identical to physical events.

The first solution has not attracted many supporters for the simple reason that it involves an abandonment of monism. In contrast, the second view is quite common. Many have thought that one should not describe thoughts and feelings as *causes* of our actions, but as *reasons* for our actions. Cause and reasons, on this view, are completely different things, and the concept of a cause in quite simply irrelevant for the description of human *actions* and their connection to our thoughts, opinions, motives and feelings. The problem with this view is that the reasons for our actions become irrelevant for the execution of them; for, if reasons and causes are different things, and if there are physical causes of the observable, behavioural component of our actions, then our reasons do not determine our actions over and above their determination by the prevailing physical. Reasons become redundant.

The third view is that mental events and states are identical to physical events and states where the reasons, i.e., motives and beliefs we propose for our actions, are all physical states of the brain. It may be that one cannot exactly pinpoint *which* physical events are identical to which mental states, but this is not a decisive counter-argument against the idea that mental and physical events are identical, either at the type or token level.

The weaker stance is token identity; it says that every *individual* mental event is identical to some *individual* physical event, but it does not follow that a certain *type* of mental event can be identified with a certain *type* of physical event. If, on a number of different occasions, Lisa is jealous of Karl, then these instances of jealousy are different individual events of the same mental type. It is conceivable that every such individual mental state is identical to some physical state in her brain without every such brain state belonging o the same general type. Indeed, describing Lisa as being jealous at different occasions may correspond to different types of brain states on these different occasions. If such is the case, then reductionism is impossible, since a successful reduction would require psycho-physical laws connecting mental event *types* to physical event *types*. Thus one can be a proponent of the view that tokens of mental events are identical to tokens of physical events and at the same time claim that a reduction of types of mental events to types of physical events is impossible. According to this view, one can also claim that a mental event is the cause of a physical event, and vice versa. For, if the mental event m is identical to a physical event f (i.e. $m = f$), and f is a cause of another physical event g, then m is a cause of g.

This is the core idea in Donald Davidson's *anomalous monism*. He strongly argues for the claim that the reduction of mental events to physical events is impossible by arguing that there cannot be any strict psycho-physical laws. Davidson admits that one can formulate approximate psychophysical laws, but he thinks this insufficient for achieving a successful reduction. A full discussion of the details of his argument will not be given here; only some of the highlights will be mentioned by way of justifying his non-reductive identity thesis.

One of Davidson's points is that one can distinguish between stated/believed motives and actual motives for actions. People in general assume that the motives for their own actions are good and known to them, and yet others often have strong suspicions that the true motives are quite different from those the agent would give. According to one of psychoanalysis' fundamental assumptions, which most people accept even though they are sceptical about much else in psychoanalysis, many of our psychological states are unconscious. So it is quite possible that an agent's officially stated motive for her action need not be the cause of the action, even if the agent honestly believe so. It is the real motive of an action that is identical to the cause of the action; that is, that motive which actually brought about the action, and nothing else.

12.8 Speculations

One way of trying to understand the relation between the mental and the material, whilst at the same time take into account the distinctive aspects of mental processes, is to characterize the surrounding debate as two different perspectives on the same object. When we use concepts like neuron, brain state and other biological concepts, we assume a third-person, scientific, perspective where the criteria for using terms can be made intersubjectively verifiable. On the other hand, when we use concepts like sensation, emotion and thought, we assume a first-person perspective. Thoughts, feelings and sensation are brain states described, using mental vocabulary, by the subject to whom those brain states belong. It is clear, by definition, that this is a subjective perspective in which we associate a qualitative subjective experience with each of the terms, a subjective experience which cannot be expressed in a third-person perspective. Wittgenstein's argument, referred to above, shows that no intersubjective criteria for these experiences can be given. Unfortunately, common language often is not clear about this distinction.

One way of understanding at least the intentional aspect of sensations is to relate this aspect to the brain's ability for sorting external stimuli. Everyone has experienced situations in which our attention is so directed towards something (a thought, a feeling or some event) that we are more or less oblivious of anything else going on in our immediate surroundings. This is an extreme situation, but such situations do often occur, though perhaps to a lesser degree. The brain's first task is always to filter the signals coming in through our sensory organs. If we did not have this ability, we would not be able to concentrate our attention on any one thing; we would be, in effect, helpless. For example, when we say that we observe a certain brand of car, or a characteristic smell, we are experiencing an effect of the brain's filtration of external stimuli. Those stimuli that remain make their way to the prefrontal cortex, the centre of consciousness, where they somehow are processed to become a perceptual object, i.e., a representation of the object towards which we direct our attention.

This way of thinking yields, in my opinion, a plausible explanation of the directedness of sensations in physiological terms. However, it does leave one fundamental question unanswered: what more is required for one to be conscious of something? Do various signals have to have certain properties, or do they simply have to reach a particular part of the brain? And what about intentional objects that do not exist?

12.9 The Science of Man

The expression 'science of man' was used by eighteenth century philosophers as referring to studies of human thinking, feeling and deliberating (The word 'man' should here not be understood as contrast to 'woman' but to 'nature'.). It is closely connected to philosophy of mind. The core problem in philosophy of mind is in my view how to integrate descriptions of the mind into a biological description of humans. I'm convinced that the mind, or the soul, is not a separate substance different from the human body. The mind-body relation is not a relation between two different things, but a relation, or perhaps a system of relations between two different ways of *describing* states of affairs and events within a human being. When we say about a certain person that she has a certain thought, desire, or emotion, we apply a mental predicate and say that the person we are talking about satisfies this predicate (See Appendix). When we say about this person that her central nervous system is in a certain physiological state, we say that she satisfies a biological predicate. Hence, when trying to understand the relation between mind and body we should study how we use mental and biological predicates to sort individual person's states into categories and how mental and biological predicates are related.

There is little hope of finding that a certain mental predicate has the same extension as a certain biological predicate, because we apply rationality principles when using mental predicates, as Davidson has shown (see Sect. 12.7), whereas no such principles are operative when we apply biological predicates. Hence, reduction of the mental to the biological seems, in my view, utterly implausible.

If our goal with the study of science of man is to gain better understanding, we should use mental predicates in our studies, because 'understanding' means knowing what beliefs, desires and emotions people has. Many hold that it is difficult, some say impossible, to formulate reliable laws, or regularities useful for predictions, using mental predicates. This stronger stance is taken by Davidson, who argues against the possibility of strict psycho-physical laws connecting the mental to the physical.

But I see a way out here, inspired by Freud, the inventor of psychoanalysis. Freud had patients whose actions appeared irrational in the sense that they did not seem to follow from the agent's conscious desires and beliefs. Freud's idea was that their actions in fact was driven by unconscious beliefs and desires, they were hidden from the agent's own consciousness by a kind of inner censor. Hence introspection into our own mind is not at all reliable; we often fail to obtain knowledge about our own desires and beliefs (Psychoanalysis is a highly controversial theory, but the notion that one may have unconscious desires and beliefs is generally accepted.). This was, by the way, the same conclusion James Watson (1913), the founder of behaviourism, drew from the complete failure of empirical psychology during nineteenth century. So if we want to retain the belief-desire-action model of understanding humans, we cannot rely on individual subject's reports about their own beliefs and desires. But another strategy is possible. Suppose we accept as a fundamental law: If an agent desires the goal G and has the belief that the best action for achieving the goal G is to perform the action A, then the agent will do A (As I argued in Chap. 10, a general statement may be a law, in spite of the fact that it is, or is part of, an implicit definition of a term used in the formulation of that law.).

This law can easily be refuted if we identify the goal, the action and the belief independently of each other. But all three concepts are intentional, so it is obvious that neither can be directly observed; identifying a particular desire, belief or action must be done indirectly, via interpretations of observable behaviour, such as responses to stimuli, answers to questions etc. Just as we identify a certain charge by observing the motion of a physical body, we may identify a certain belief or desire by observing the outward aspect of the agent's action, his behaviour. It might be more complicated in the mental domain than in electromagnetism, but I see no principal difference. The crucial restriction is that the agent's own reports about his/her desires or beliefs should not *automatically* count as evidence against such an attribution, just because such reports are the results of introspection. Introspection is unreliable.

Daniel Kahneman's book, '*Thinking fast and slow*', are full of such examples (See more about this book in Sect. 14.1). If, for example, the fast system determines our action, we can describe its operation as that an unconscious belief and desire is formed such that the appropriate action takes place. One could perhaps object by saying that unconscious mind-states cannot be beliefs or desires; such states are by nature conscious. But, again, we have strong evidence that we are not conscious of most of what goes on in our minds; this was a main reason for Freud to invent his theory about unconscious dynamical processes. One could be sceptical about Freud's theory without denying that the distinction between conscious and unconscious mental states is a distinction with a difference.

In fact, the concepts of belief and desire, in ordinary parlance, has a strong functional character; they are names for those states of our minds that guide our actions; and calling a state a belief state doesn't entail anything about its mode of operation in producing an action, or that it is conscious. So I see no fundamental difficulty in a science of man aiming both at understanding and at prediction.

Many argue that man and society is so much more complex than simple physical or chemical systems so that there is no hope of making reliable predictions. But the premise that man and society is more complex I outrightly reject. Most purely physical, chemical and biological systems have an enormous complexity, as measured for example by the number of degrees of freedom. I see no difference in that respect between a human being, a human being in its environment, or a society and a system studied by natural science.

A second argument for the difference between science of man and of nature is that for ethical reasons we cannot perform controlled experiments in strictly controlled circumstances. This is true, of course, but the same problem is prevalent also in many areas of natural sciences. One way to overcome this is to study partial systems and then construct models for their combination; and this can be done also in the study of man. Another is to use statistical methods, which, as we have seen, gives *some* indications about possible mechanisms and causal structures. And this can be done also in the study of man.

A third argument for the difference between the study of man and of nature is that humans react and change their minds when being studied. This is certainly true. Humans are essentially agents whose beliefs, desires and volitions may change when being described by others, whereas things studied by physics, chemistry or biology are not in this sense agents, they have no thoughts. But I doubt if this really is a difference that have methodological relevance. The fact that people are affected and may change their minds when being subject to actions or descriptions from other people is a typical example of a feedback mechanism. And the same is true in micro physics; observing an object means exchanging photons, i.e. portions of energy with it, and when observing the smallest objects they may change their state, causing difficulties in making predictions. The mechanisms differ, but the fundamental methodological difficulty, to make reliable predictions, is the same.

Still, both in physics and in the study of man we are sometimes able to make predictions of the evolution of ensembles of individuals, particles in the case of physics and people in the case of the study of man.

All in all, I see no fundamental obstacle for future progress in the study of man using mental predicates in the descriptions: And by progress I mean not only to provide understanding of individuals or groups of individuals and societies, but also results with predictive force, at least predictions in terms of probabilities.

Discussion Questions

1. In the film *Alien*, there is a scene where two crewmembers of a spaceship become violently disagreeable and begin to fight. During the fight, one of the crewmembers falls against a sharp corner. To everyone's surprise, the crewmember's wound does not release blood but hydraulic oil! This 'crewmember' was in fact a robot, which the villains had smuggled onto the ship in order to destroy it.

Suppose that this robot behaves just like a human; that it speaks, jokes, eats and so forth. Is it possible that this robot lacks consciousness; that it has no feelings, thoughts or sensations? If your answer is yes, what more do you require of the robot for it to be capable of these experiences?

2. Within computer linguistics, researchers are eagerly investigating technology that can translate one natural language into another, and much headway has been made. In the E.U., for example, computers are now responsible for the brunt of the work in the translation of official documents. It does not seem unreasonable to believe that computers of the future will be able to translate non-fictional texts just as well as a human. Would we then say that such a computers understand the meaning of various words and expressions? If not, why would we say that a human who performs the same task does understand the meaning of the text he or she translates?

Further Reading

Chalmers, D. (1996). *The conscious mind*. Oxford: Oxford University Press.
Churchland, P. M. (1988). *Matter and consciousness*. Cambridge, MA: MIT Press.
Churchland, P. M. (1995). *The engine of reason, the seat of the soul*. Cambridge, MA: MIT Press.
Churchland, P. S. (2013). *Touching a nerve. Our brains, our selves*. New York: W.W. Northon & Co.
Davidson, D. (1980). *Essays on actions and events*. Oxford: Clarendon.
Dennett, D. (1981). *Brainstorms*. Cambridge, MA: MIT Press.
Kim, J. (1998). *Philosophy of mind*. Boulder: Westview Press.
Putnam, H. (1960). *Minds and machines*. Reprinted in Putnam 1975, pp. 362–385.
Putnam, H. (1967). *The nature of mental states*, reprinted in Putnam 1975, pp. 429–440.
Putnam, H. (1975). *Mind, language, and reality*. Cambridge: Cambridge University Press.
Searle, J. (1992). *The rediscovery of mind*. Cambridge, MA: MIT Press.
Watson, J. B. (1913). Psychology as the behaviourist views it. *Psychological Review, 20*, 158–177.
Wittgenstein, L. (1953). *Philosophical investigations*. Oxford: Blackwell.

Chapter 13
Science and Values

Where I benefit, that is where I place my faith.

Proverb

13.1 Values and Their Role in Science

According to a common norm, science should be free of values. This view is based on the idea that values are subjective, whereas science strives for objectivity. Today, the view that all valuation is subjective is quite common, but such has not always been the case. The default view has been that moral and aesthetic judgements are true or false; their content are propositions about moral or aesthetic facts. If one accepts this traditional view, then the motive for claiming that science ought to be value-free disappears. For why, then, would one dismiss certain kinds of descriptive propositions as not allowed in science? However, if one assumes that value statements are not descriptions of objective conditions but rather expressions of the speaker's feelings and subjective attitudes, then the requirement that science is value-free is extremely well motivated. The view that value statements are not really statements but rather expressions for our feelings is called emotivism.

The norm that science should be value-free can be interpreted in various ways. A strong interpretation would be that science should not contain any expression of values. This is not reasonable since one would perhaps want to investigate which values there are and what effects they have on the actions of various groups. Such research cannot be said to break with the norm above. A more reasonable interpretation is that the individual researcher should not express any of his/her own values. However, one can argue that there could be a researcher who unconditionally gives a complete and accurate description of the consequences of various measures (where, for example, a politically relevant issue is being discussed) and concludes by explicitly recommending a particular measure, based partly on the researcher's analysis of the aforementioned consequences, and partly on a clear expression of the value premises. For the outside reader, it is easy to discern what in this text are scientific results and what constitutes valuations. I can find no reason to object to such an exposition.

L.-G. Johansson, *Philosophy of Science for Scientists*, Springer Undergraduate Texts in Philosophy, DOI 10.1007/978-3-319-26551-3_13

Upon reflection on this imagined case one is lead to formulate a third interpretation of the norm that science should be value-free: *one ought not to deduce value statements from premises that only contain descriptive statements.*

This is often shorted to the slogan 'one cannot deduce an ought from an is' was first claimed by Hume, and therefore often called 'Hume's law', has much to say for it; for example, that it is a consequence of our concept of valid inference. Nevertheless, we often find argumentation that breaks this rule. Consider the following example, which mirrors the sentiments now (2015) among lots of Swedes:

> The number of immigrants to Sweden has risen significantly the last years. Therefore we should impose stronger conditions for immigrants to Sweden.

That one often makes this jump from the descriptive to the normative in arguing about political issues is the result of leaving out the intermediate premise (in this case the premise is 'the increase in immigration is undesirable').

Political argumentation, like all use of language, always occurs in a certain context. The contextual factors help to supplement the message, allowing the author of that message to treat various things as implicit. The author does not need to say that which can be presumed to be well known or obvious for their audience. In fact, it does not even have to be that obvious for their audience, as presuming certain things to be obvious can be a rhetorical trick employed to increase the weight of ones arguments.

The difference between common and more scientific argumentation is, among other things, that a scientific argument places greater requirements on the rigor and clarity of ones premises. This general norm leads to the more precise norm that one should not leave implicit value premises in scientific argumentation. Leaving out such a normative premise in an argument results, just as in the example above, to an argument breaks with the rule of not inferring an ought-statement from only descriptive premises.

However, the application of this requirement is complicated by the fact that there are many words and expressions that are neither purely descriptive nor purely evaluative. Some examples are words that describe human character traits: 'uncompromising', 'inflexible', 'light-hearted', 'thoughtful', 'indecisive', etc. For every human character trait, it is possible to find two (or more) words, one positive and one negative, such that no matter which we choose we get some measure of valuation. Another group of examples can be found in political language: 'freedom', 'democracy', 'prosperity', 'society', etc. It is hardly possible in practice to avoid such descriptive and yet normatively charged words in scientific accounts within, for example, political science, economics and psychology. Therefore, the norm that one should not deduce normative conclusions from descriptive statements is not so easy to follow. One way to be as clear as possible is by being extremely explicit in what one packs into these concepts and, for example, by distancing oneself from eventual evaluative components where appropriate.

Most, but not all, philosophers basically agree with Hume in distinguishing between facts and values. Hilary Putnam is one who disagrees, see his (2002).

13.2 Value-Free and Value-Laden

It is sometimes claimed that research is not value-free if it is funded and governed by economic, political or religious groups. The argument is invalid since it builds upon a confusion of 'not value-laden' and 'value-free', an important distinction once introduced by Weber.[1] That research is value-laden means that the governance of research, the choice of which questions that is be the subject of scientific investigation, is driven by goals that are chosen with certain values in mind. That research is value-laden is thus something completely different from not being valuation-free. Research aimed at finding better treatment for diseases are certainly value-laden; we would give very high value to positive results in this area. But, of course, reports of such results, if they are found, should be value-free; we want objective knowledge about what to do for curing and/or preventing severe diseases. That is to say, scientific activity can be value-laden and value-free at the same time.

In principle, one cannot dismiss a scientific result with the argument that the researcher has been paid by someone with suspicious interests. If one does so, then what you have is an example of an *ad hominem* argument: the objection to the view is based on the person who presents the view and this person's hidden agenda, and not on the merits of the argument.

However, in practice it is often the case that we are suspicious of results presented by a researcher who produces results that are favourable to the researcher's employers. Such is especially the case as regards politically charged questions about economic politics, environmental politics, social politics, and so on. A clear example of this is the discussion regarding the economic development in Sweden during the period 1970–1995. Many economists, critical of the generally social democratic politics of this period, concluded that Sweden's place among the world's richest countries has fallen from number 3 at the beginning of the 1970s, to number 18 during the mid 1990s. Furthermore, many held it confirmed that this drop is the result of the economic politics during that period, explaining this development as the result of a sort of sclerosis, a structural rigidity, in Swedish society that has resulted from too much social politics, wage equalization and high taxes. But certain social democratic economists, especially Walter Korpi objected to this analysis; in particular, he objected to the factual claim that Sweden lagged behind in the development of prosperity. He has objected to the period chosen for measurement, the choice of countries used in the comparison and the analysis of the causes cited in the argument. For an outsider, it is naturally difficult to have a grounded view of this matter, but it is pretty clear that there are numerous judgments intertwined within these economic reports; judgments concerning the relevant measures, selection criteria, objects for comparisons, etc. It is naturally

[1] See Weber (2011). For a thorough discussion of Weber's position on this matter, see Bruun (2007).

extremely difficult not to be influenced by political valuations with respect to such judgments. However, just as one may be sceptical about statements made by, for example, corporate economists, since the corporation in question could have its own interests in changing Swedish politics, oné can equally be sceptical of Walter Korpi's statements, since he has a strong ideological interest in showing that the social democratic politics of the day did not have long term negative effects on economic development. For if such were that case, then one would have to admit that the crucial parts of the political system were a failure and that the bourgeois critique was correct (When this translation is finished, 2015, the politics has been changed in some of these respects, both by social-democrat and conservative-liberal governments, and the Swedish economy has regained productivity to a considerable degree. This indicates that the 'sclerosis diagnosis' at least to some extent was correct.).

In summary, one can say that the fact that a certain research effort is funded by a proprietor who has a strong interest in certain conclusions, or that the results in some way stand in relation to the researcher's non-scientific interests, provide some reason for suspicion. Conversely, if a researcher comes to a conclusion that undermines the researcher's, or his proprietor's, economic or ideological interests, then there is even more reason to believe that the researcher has been very careful in her attempts at validating her results. However, such considerations certainly cannot be viewed as relevant arguments for, or against, a concrete proposition within a scientific debate. One must, under all circumstances, make the effort to take a stance regarding the matter itself. Regarding the concrete issue discussed above, concerning economic policy in Sweden, I have no competence to judge the matter.

13.3 Is Science Valuable?

There is almost universal consensus concerning the value of science and that it is a good thing to spend some of the taxpayers' money on research. Many researchers think that one should spend still more money on research, but one could suspect self-interest to bias them on this issue, but virtually no one opposes spending *some* money on research. But there is substantial disagreement concerning what kind of research should be supported and what fields of research should be pursued.

But let us proceed to a general perspective; how should we quite generally motivate spending taxpayers' money on research? The answer seems obvious; in the long run taxpayers will benefit from the results, for example by new, more effective medical treatments and more energy-effective engines being developed. Of course, it is not just a case of issuing an order to the scientific community; there are no guarantees that a particular research project will succeed. Nevertheless, it seems rational to spend money on a spread of research projects in the hope that some will be successful. A politician responsible for funding research could argue just like a venture capitalist risking his money in quite a number of companies and projects. The venture capitalist is well aware of the high risks and is prepared for failure of most of the projects they support. But he hopes that some will succeed, and that the return on those successes will be more than sufficient to compensate him for the failures. In the same way a research policy maker should be prepared for quite a number of projects failing, but hoping that some will succeed in making a discovery that is so important that it more than fully compensates for all the failures.

Everything so far said presupposes that the point of doing research is that some results will prove useful. But what is useful? How do we determine that?

When it concerns medical science and technology, broadly conceived, one could say that useful results are expressed as propositions that have the form of law-like generalizations, i.e. propositions that can be used for predictions and goal-directed actions; in other words, knowledge about general causal connections, and the general aim is mostly to find such law-like generalisations. Part of social science and the humanities also have this aim, but great parts of social science and humanities cannot conceivably be useful in this sense. Most research in sociology, history, archaeology or philosophy does not aim at producing law-like and applicable generalizations. Also mathematics and foundational research in theoretical physics may not be useful in any obvious sense or to any timescale that would warrant the investment they require. How can such research be motivated?

Quite often defenders of research in humanities say that they are useful in a broader sense; the result may improve our quality of life. I don't think it is a good strategy using the word 'useful' for defending research in humanities. The reason is that this defence glosses over the distinction between things that are useful means to an end and things that are valuable in themselves, and the default meaning of 'useful' is as a means to an end. If an historian argues that research in history is useful, a sceptic taxpayer could say: 'Useful for what?' (Some might defend research in history with the argument that we can learn from history and avoid earlier mistakes; but it is doubtful whether different events in history are sufficiently similar to enable useful generalisations.)

Another answer, once formulated by Max Weber, is that the aim of research in humanities and social sciences (German 'Kulturwissenschaften') is to provide *understanding*, i.e., understanding of man, culture and history. Understanding of these phenomena means understanding the meaning of phenomena, i.e., understanding which beliefs, emotions and motives that are connected with these phenomena. Hence the goal of research in social sciences and humanities is not instrumental; these activities are not directed toward aiding us in achieving some end/goal. Rather, these activities, and the understanding they engender, are goals in themselves, it is simply things that interest many people.

Philosophy, at least some part of philosophy, mathematics and perhaps some other disciplines may be said to form a third category of research. Inquiries in these areas are not useful in any direct sense, nor do they provide understanding of man and culture in the sense indicated above. These disciplines are best described as critical scrutiny of our thinking and of development of concepts. Concepts are our vehicles for thinking and philosophy and mathematics are concerned with some very fundamental concepts that can be used in virtually all other disciplines (think of concepts like *knowledge, evidence, proof, function, relation*). And this may be defended as being useful in the very long and very broad perspective.

If we say that some activity is valuable because it is useful in the sense of being a means to an end, we are only postponing the question of why it is valuable, for why do we value the end result? Why is it good to cure diseases, diminish famine, increase productivity or improve our thinking? Obviously, we come very quickly to things where we will say 'This is good in itself.' So the conclusion is that on this matter all scientists are on an equal footing when asking for money from the taxpayers. The basic issue is what people want, what they think is valuable in itself.

People value different things and limited resources must somehow be distributed. The usual procedure of research funding is a two-step process. In the first step representatives for the public determine the general policy (more money on new forms of energy production, or more money on better crops?) and in the second step representatives for the researchers determine which projects to support in a specific field. The idea is that the first decision is not a scientific decision; it is a value-driven choice, whereas the second decision is expected to be better if made by experts in the field, for they know which projects are more likely to succeed on delivering the policy goal. But some are suspicious that the selection procedures are not better than a purely random procedure.

13.4 Feminist Critique: Hidden Values in Science

In later years, feminists have criticised certain research because it is 'sexist'; that is, it contains more or less hidden assumptions that imply a devaluation of women. A clear example is the research surrounding myocardial infarction, which is mainly done on men. We now understand that myocardial infarction is common even among women, but the symptoms may be different in women than in men. This has resulted in an under diagnosis of infarction in women and also in faulty knowledge of how it should be treated, since the medicines that exists are not tested on women. It has been claimed that this practically implies a devaluation, or neglect, of women's problems.

Another, slightly deeper, form of male narrow-mindedness in science is when concepts, which are based on cultural phenomena in today's society, are used as naturally determined, universal categories. A clear example is when the current division of labour between the sexes (that men generally engage in outgoing activities, achievements, careers, competition in the workplace, etc., while women focus on children, homes and caretaking) is taken as a starting point for describing social patterns in prehistory. A common account of prehistoric human development is the following: Our ancestors developed four features simultaneously and in intimate interaction: one began to use tools, one began to walk upright, one began to hunt, and the size of the brain (and hence energy consumption) increased significantly. A natural conclusion from this is that the use of tools made more efficient hunting possible, which was required to support the larger brain. Findings of primitive tools, which are assumed to have been used for killing larger animals, support these assumptions. Thus one is presented with a picture of typical male behaviour that has driven human development forward. Contrary to this, firstly, it has been claimed that the primitive tools that have been found–differently shaped stones–cannot be unequivocally connected to hunting; they could just as well have been used to work with plants and roots, or something else 'typically female'. Secondly, one can argue that the tools needed to gather vegetables for eating were generally made of less durable materials such as wooden sticks, and therefore there is no evidence left that shows how important the collection of vegetables was for our ancestors. This claim is supported by studies of present-day people that live as hunter-gatherers, surviving mainly by collecting fruit, nuts, roots and other parts of plants (which is regarded as typically female activities). Thirdly, how do we know that the gathering of plants was mainly a female activity? In short, there is a strong tendency to project the current state of society on completely different cultures by using categories for sorting data that are heavily influenced by the society we live in.

The main point of this critique seems not to be that the scientific results are false, pure and simple, but that focusing on certain circumstances and neglecting others is driven more or less by unconscious norms about what is 'normal' male and female behaviour. One should notice that the word 'normal' is ambiguous. It can mean 'common' without any valuation, but also 'natural and right', which clearly has evaluative content. Often one shifts between these two meanings without noticing it.

When we describe our empirical data, we must use general concepts that we as humans have constructed. But we humans are rarely, or perhaps never, neutral observers of the external world. That is to say, we are first and foremost agents who often want to understand the world in order to intervene and influence it. So we form our concepts, often unconsciously, as a part of our desire to shape the world or defend a certain way of life. This normative component of the conceptual framework comes to be known as a valuation of what is important, relevant or essential to a complex phenomenon. The general conclusion one can draw is that the categorization of phenomena is interest-related. This connection to interests is not something one can get rid of, but one can request that researchers be aware of what purpose a certain categorization has.

One possible objection to the above is that science should make *correct* categorizations of phenomena; scientists should describe things 'as they actually are'. Even if there is a purpose, it is secondary. For if one has categorized a phenomenon in a way that does not agree with reality, then it is quite simply false, regardless of its purpose. This objection is based on a kind of essentialism; things have essential properties, and it is the task of science to discover them. However, I am quite sceptical of this view; I share Quine's hostility towards the distinction between essential and contingent properties of things, objects, events or states of affairs, for the same reasons as he gave.[2] Unfortunately, further discussion of this complicated, but rewarding, topic lies far beyond the scope of this book.

13.5 Research Ethics

Doing research quite often means taking a moral stance, balancing the value of the possible results against the interest, dignity and integrity of humans involved. Such balancing has not always been made. An infamous example of research, which triggered heavy moral critique, was the Tuskagee clinical study conducted between 1932 and 1972 by the US Public Health Service to study the natural progression of untreated syphilis in African-American men in Alabama. These men were told that they were receiving free health care from the U.S. government and those who in the study group who had syphilis, 400 out of 600, were not told about their malady. In 1940 it was established that penicillin was an effective treatment of syphilis. One would then have expected that the researchers should have ended the study and given all persons in the study group with syphilis the new effective treatment, but they did not, neither did they inform them about their disease and its proper treatment. The study was finally terminated in 1972, as a result of a whistleblower reported to the press what was going on and 25 years later, in 1997, President Clinton formally apologized and held a ceremony at the White House for surviving Tuskegee study participants. He said:

> What was done cannot be undone. But we can end the silence. We can stop turning our heads away. We can look at you in the eye and finally say on behalf of the American people, what the United States government did was shameful, and I am sorry ... To our African American citizens, I am sorry that your federal government orchestrated a study so clearly racist.

This and other cases of morally bad research have resulted in ethical questions related to research have come increasingly to the fore.

Another example of research, not so malevolent as the Tuskagee study but still morally doubtful, was Milgram's well-known experiment for studying obedience. A number of volunteers were asked to help in carrying out a learning experiment. These people would give other volunteers certain tasks, and if the latter failed the task, the former were told to give the latter electric shocks. For each task the volunteers failed, they would receive more powerful shocks. The underlying hypothesis was said to be that punishment could be used to improve learning. In reality, this was a bluff. When the test subjects flipped the switch, no electric shock was administered, and the person who was supposed to be undergoing the 'learning experiment' was instructed to pretend to feel a shock. Many of the participants went

---

[2] Quine's hostiliy to the distinction between essential and contingent properties is that it depends on how we describe things and most things can be described in many ways. So the distinction is not objective, it varies depending on which the description the speaker thinks most appropriate.

quite far and, as they believed, 'increased the intensity of the shock' more and more, even though the 'test subjects' on the other end signalled that they were in heavy pain. The point of the experiment was to see how far these volunteers were prepared to go in causing other people pain if they were instructed to do so by a person in a position of authority, i.e. the researcher (The sad result of this experiment is that most participants obeyed even though they believed that they were hurting another person quite severely.).

One generally accepted norm is that all research where humans are involved should only be carried out with the participants' informed consent. This means that one must inform the participants of the purpose of the research and acquire their consent to participate. Obviously, Milgram's experiment strode against this norm. However, Milgram could defend his actions with the following argument: firstly, it would have been impossible to perform the experiment if the test subjects had known the true purpose of the experiment, and it is important to study people's propensity to obey orders even when the people in question believe that doing so will cause others pain. Secondly, the test subjects were informed after the experiment that these other people had not been harmed. In response to Milgram's defence one could argue that it is doubtful if one can draw any general conclusions from this laboratory experiment, and that it is degrading to be tricked as the volunteer test subjects had been. This second concern is especially pertinent for those who were inclined to give their 'victims' powerful shocks since they appeared to the researchers to be quite unsympathetic people.

Another ethical problem in research sometimes arises in the testing of a new medical treatment. It is often the case that one performs a preliminary control test after some time has passed to see if there is any difference in the tendencies of the tested treatment and the treatment that the patients in the control group received. One example is the testing of AZT to AIDS patients. In a follow-up, it was found that the patients that had received AZT were in much better condition than the control group, and so it was decided to end the experiment early and give all participants AZT, even those who had been given an ineffective drug. This may seem to have been an unproblematic move, but the researchers were faced with an important decision. Every general conclusion regarding long term effects of a certain treatment drawn from a sample test is more or less uncertain, and the uncertainty is naturally increased the shorter the test lasts. One is faced, therefore, with the choice between giving all patients the treatment and sacrificing reliability, or to continue the experiment and risk denying certain patients a treatment that could have helped them. It may seem to be a rather easy choice, but one ought to consider that the preliminary conclusion, that the treatment has a positive effect, could be wrong. In such a case, were one to shut down the experiment and give the control group the treatment, one would not have helped anyone and would have lost the possibility of acquiring the more secure knowledge of a complete experiment.

In many countries research founding bodies now require approval from an ethical committee of a proposed research project involving humans as objects of study. In addition, researchers in medicine, psychology and some other fields have in many countries established ethical codes for researchers and practitioners in their respective fields. As an example of an ethical code in psychology one may consult American Psychological Association's 'Ethical Principles of psychologists and Code of Conduct', see http://www.apa.org/ethics/code/principles.pdf (In the literature list referred to as 'APA ethical code'). This document states five general principles:

Principle A: Beneficence and Nonmaleficence.
Principle B: Fidelity and Responsibility.
Principle C: Integrity.
Principle D: Justice.
Principle E: Respect for People's Rights and Dignity.

For a thorough explanation of what these principles means in practice see the document.

Further Reading

APA-code: http://www.apa.org/ethics/code/principles.pdf
Bruun, H. H. (2007). *Science, values and politics in Max Weber's methodology: New expanded edition*. Aldershot: Ashgate.
Harding, S. (1991). *Whose science? Whose knowledge?* Ithaca: Cornell University Press.
Longino, H. (1990). *Science as social knowledge*. Princeton: Princeton University Press.
Merton, R. (1973). The normative structure of science. In *The sociology of science* (pp. 267–278). Chicago: University of Chicago Press.
Weber, M. (2011). *Methodology of social sciences*. New Brunswick: Transaction Publishers.

Chapter 14
Some Recent Trends in Science

14.1 The Impact of University Mass Education

Since antiquity and until rather recently very few people were occupied with what we now call science, i.e. systematic thinking about nature, society and humans. This has changed profoundly the last decades in all western societies, and is on its way in the entire world. This is, one may assume, related to the expansion of higher education. Some 50 years ago, relatively few people, less than 5 % in advanced countries had a university degree, while nowadays it is 40–50 %. One may perhaps assume that most of what students learn during their academic studies is based on science and hopefully approximately true. Hence one may entertain the hope that greater and greater portions on mankind are able to make better decisions, act cleverer and not so prone to devastating mistakes as in older times. I do entertain that hope, albeit there is also reason for pessimism.

One such reason may be found in research into human decision-making, as reported by the Nobel laureate Daniel Kahneman in his book *Thinking: fast and slow*. In this wonderful book a huge number of experiments on human thinking is discussed. Kahneman's core message is that we have two systems for decision-making, the fast system and the slow system. The fast system acts unconsciously, using inherited and early in life learned emotional responses: good or bad for me? Thus, for example, new faces immediately give us a positive or negative feeling which guides our actions. Sometimes this fast decision system is very useful, even life saving, as when we feel a strong impulse to run away from a dangerous situation. Kahneman's analysis of this system is that it often uses rules of thumb and association circuits for producing a quick response. But association mechanisms sometimes produce clearly irrational reactions, via a mechanism called priming.

The other system is the slow one, where we deliberate, consider alternative actions, estimate how reliable the available information is etc., in short when we use our rational faculty. The most interesting result reported in the book is, in my view,

L.-G. Johansson, *Philosophy of Science for Scientists*, Springer Undergraduate Texts in Philosophy, DOI 10.1007/978-3-319-26551-3_14

that the fast system often makes the decisions, while the agent himself believe he has considered the alternatives and can give good reasons for his choice. The conclusion to draw is that we often think we are more rational than we in fact are.

It is possible to improve ones rationality, and to train oneself to be a bit sceptical against ones 'gut feelings' in situations where there is some time do make a decision. On the other hand, 'gut feelings' may be the result of unconscious reception of information, of for example other people's attitudes and emotions, which they try to hide. So when should one succumb to the first thought, to ones gut feelings? No easy answer is available, so far as I know. Nevertheless, I think one may entertain some hope that the future for mankind will improve because of deeper and more widely disseminated scientific knowledge.

All in all, there is, I think, *some* hope that more widespread scientific knowledge will improve the conditions for mankind, and for many individuals. And in democratic societies one may hope that policies will be cleverer if the general level of scientific education improves.

14.2 Publish or Perish: The Value of a Research Paper

Not so long ago (100 years?) nearly all scientists were either wealthy men (seldom women) doing science as a hobby, or university teachers who primarily was hired for teaching while doing research in their spare time. This has changed. Very few people do research without being paid for doing just that and universities now select professors for hire based primarily on their track record as researchers (although most have a non-negligible teaching load). Professors all over the world now see their prime duty as doing research. And since the enrolment to university and college studies has risen tremendously in the developed world, a huge number of university teachers have been hired. All these people compete for better positions and promotion, and since the main merit in academia is research results in terms of scientific publications, they focus efforts mostly on publishing scientific papers, not on their teaching. So, because of the enormous increase in students at universities there has been an enormous increase in university teachers and thus in the number of published papers. Hence this increase is to a considerable extent a side effect of the increase of basic academic education.

Scientific papers must be read in order to be of any value. Alas, there is strong evidence that the great majority of scientific papers are never read by anyone else than the referees and the journal editor.

First of all the majority of papers are never cited, and secondly, there are findings (see for example http://www.complex-systems.com/pdf/14-3-5.pdf) suggesting that only circa 20 % of cited papers are actually read by the citer (This paper I have read!). One may conclude that only when a paper are cited five times or more can we have some confidence that it actually have been read.

One could of course argue that many papers are read without the reader finding reason to quote it. This is plausibly true, but it only strengthens the hypothesis that, from the mankind's point of view, most scientific papers are of no worth. For if a scientist read a paper but don't have reason to refer to it in his future research, it had plausibly no or little impact on this researcher's thinking.

The researcher's subjective aim of publishing a paper is his/her own career, as already pointed out. But why not putting more effort in writing for a broader audience, thus promoting the general level of scientific understanding in society? From a societal point of view it would be clearly rational, but, alas, from the point of view of the individual researcher it is not, since writing for a broader audience is not valued in academia. As Nicholas Kristof wrote (New York Times, February 15, 2014): 'If the *sine qua non* for academic success is peer-reviewed publications, then academics who 'waste their time' writing for the masses will be penalized.'

One may conclude that in most cases writing a scientific paper is waste of time. But there are, of course exceptions, some papers are indeed highly valuable.

14.3 Research Funding and Planning

Could one diminish waste of money and time on bad ideas and increase the proportion of useful research? I'm sceptical. Research funding bodies must act just like venture capitalists. Such capitalists, I have been told, select a number of projects/companies and put in their money in these. They know that most projects are bound to fail, but since it is impossible to know which, they have to spread the risks and a reasonable number of projects are supported. Hopefully one or two succeed so that the return on that investment compensate for the losses on the failed ones. The decision problem for research funding bodies is similar and they should apply a similar strategy in order to optimize the value of the research done. I see no other rational option.

However, those funding research naturally would like to get some guidance on which projects to support. One could for example look at history of science to see which kind of strategies have worked in the past. One such idea is give money to researchers that have a good track record, people who haven proven their ability to produce relevant new scientific knowledge. Several research councils have adopted this strategy of giving money to excellent researchers. Such a strategy will result in strong bias for senior, or very senior researchers. Is that a good idea? Wouldn't it be much better to give money to newcomers, who have new ideas?

The problem of deciding which projects to support and which hypotheses to test is the question about which inductive inferences to give credence. But this is an unsolvable problem. It is impossible to give a general justification for inductive reasoning and a fortiori for selecting one project among a number of alternatives that objectively is the most probable one to succeed.[1] There cannot be any way of

[1] To use Bayesian reasoning to update one's credence is no solution, since one must have a prior probability as a starting point and no objective way of finding that is available.

sorting out those inductive conclusions that later will be confirmed and those that will be falsified, and hence no good reason to invest resources in one project rather than another. As Quine once remarked: 'The human predicament is the Humean predicament' (Quine 1969, p. 72). Hence it is no good idea to concentrate too much resources in any particular research programme, whatever the arguments. Who is to know that the chosen line of research is not a blind alley?

My conclusion is that the great scientific achievements the last decades are explained by the enormous increase in money for research and in number of researchers. It is like mass bombing from high altitudes of an area: if the bombing is heavy enough, some bombs will hit the targets. More bombs, more targets being hit.

14.4 Big Science

Big science began in physics with the construction of big particle accelerators, such as that at CERN in Geneva or SLAC at Stanford. These investments were not motivated by any desire to test any specific hypothesis, but a recognition of the fact that testing any hypothesis about the fundamental structures of matter require this kind of equipment (And one motivation for the US congress' funding such research was the arms' race between Soviet Union and USA.). Well, one cannot, as a matter of principle, dismiss the possibility that it will be possible to probe deep into microphysics by other, cheaper methods; but so far we see no alternative.

Now we see something analogous in social science, although much less money is needed. Since the costs of collecting and storing enormous amounts of data has fallen dramatically one may consider building and maintain really big data bases, and the parallel rapid development of computers make it possible to survey these for correlations or other things. We can do data mining. Even though the costs are minute in comparison with the cost of building and running CERN, there is nevertheless a somewhat parallel phenomenon, in so far that basic research facilities have been constructed without there were any particular hypothesis up for immediate test.

Will these investments pay off? Well, I think theoretical physicists would say: CERN has given us the possibility to test the standard model; furthermore we have now confirmed beyond any reasonable doubt the Higgs mechanism, predicted 40 years earlier. This is deeply satisfying. But even the opposite result, a disconfirmation of the Higgs mechanism would have been exciting, since it had forced us to accept that there is something wrong with the standard model and new ideas would have been asked for. So from a scientific perspective the return of this investment was really good.

Some European taxpayers may think differently, asking: why should we pay for this? We can't see any benefits for us, nor that it has any public interest. CERN is a playground for academics.

I don't know whether these sentiments are common, nor can I see any effective way to rebut the argument. At bottom, this is a dispute about basic values. If someone says, in the vein of Aristotle, that man by nature is curious, we want to know no matter the usefulness of the knowledge, and another says, 'I disagree and I don't want public money to be wasted on that', I see no rational way to adjudicate this dispute objectively. Disputes about values are settled at the polling stations in democracies.

A different, but equally unsolvable problem arises when constructing big databases with all sorts of information about human beings. The costs are much smaller, and the possible benefit might be easy to argue for, but privacy of individuals may be threatened. It is common practice to take measures to protect privacy and integrity of those the data are about, but are these enough? One need to strike a balance between privacy and usefulness of the data base, and it is bound to be people disagreeing about the balance.

14.5 The Scientific Attitude and the Search for Meaning

In Chap. 1 I described the beginning of scientific thinking as an effort to try to understand events in nature as effects of natural causes, not the result of interventions of gods, spirits, demons or other creatures with desires, volitions and goals.

In the course of time the idea of natural laws developed and therewith the conception that events and states in nature, including our bodily states, are determined by initial conditions and deterministic causes. Hence, there is reason to try to predict some aspects of future using hypothetical reasoning; if I do such and so, then that will happen; but I do not want that to occur, so I had better do something else. And success has sometimes, some might even say quite often, followed.

It should be noticed that such hypothetical thinking might be useful even in the absence of any detailed knowledge about the mechanisms. Clear examples of this are observations of how the plague spreads from area to area. Although one had no knowledge about the agent, bacteria, one could do *something* to restrict its distribution. During the outburst of plague in England 1665 one took strong measures, such as putting ships from abroad in quarantine for quite some time, restrict domestic travels and closing the universities (which, by the way, made it possible for Newton to return home to his mother's farm to work on his magnum opus *Principia*).

Our possibilities to control our environment has increased tremendously due to better knowledge about laws and mechanisms. But still there are, in Quine's words, unsettling surprises. When such surprises hit us, such as severe diseases, accidents, tsunamis, etc., most of us are hard to accept that there were no purpose, no reason, and no explanation. Our desire for an explanation, which provides a sense of meaning of a personal disaster, is deeply felt and few are by nature prone simply to accept that one had bad luck.

This desire for meaning-giving explanations of the misfortunes in our life is connected to our natural desire to understand our own life and ourselves. Few humans, perhaps none, are willing to look back to their own life and view it as a sequence of events totally without meaning, without purpose. Humans by nature construct meanings in their own life. We try to integrate random events with dire consequences into some of meaningful pattern when we reflect on our life. It is not uncommon to hear people saying about such an event that it had a meaning after all; I learnt something, I became a better person, I met my husband/wife, etc. That is, even if we know that it was a random event, we give it a meaning ex post facto. Pious Christians in particular try to see a meaning also in severe suffering (it make you a better human, more humble, more compassionate). And many people, when they look back at a disaster in earlier life, which they were able to cope with in the long run, say: there was a meaning after all.

This is a remnant of the pre-scientific thinking where all events were considered as the effects of purposeful actions of spirits, gods or demons. When we are able to think of something as an agent, as something with a mind, we say that we understand it. To understand someone, is, as Collingwood puts it, 'to put oneself in the other's shoes'.

Should one denounce this urge for understanding of disasters and other unpredictable events as bad thinking? My view is that it depends on which conclusions one draws. It should be clear that calling something meaningful, in the sense here given, has no implications for our future actions and no predictive force. So if it has no impact in our actions and decisions, it seems innocent and perhaps beneficial for well-being (It might even function as placebo.). But if someone allows thoughts about the meaning of natural events to guide his choices about how to act, he runs the risk of making, from his own point of view, bad decisions.

No one can live, I grant, without trying to find a meaning in his/her own life. But we should not allow this craving for meaning to interfere with our search for reliable knowledge about laws and regularities relevant when deliberating on which actions to take. Of course, we may make wrong inductions about which regularities there are and hence which actions best to do. But trying to use wishes of gods as guidance is bound to fail.

Appendix: Logical Forms

Introduction

When we disagree with other people, we discuss. If we really try to reach for knowledge and not just to win a debate, we are faced with the question as to whether or not an opponent's argument should be accepted. In order to decide this we must answer two questions: (i) do we accept the opponent's premises, and (ii) do we accept the reasoning, or form of the argument? If the answer to both of these questions is 'yes', then one must accept the argument. If on the other hand one finds a premise untenable, one can object in the following way: 'I accept your form of reasoning, but your starting point is false, and therefore I cannot accept your conclusion'. Another possible stance is 'I accept your premises, but the conclusions you draw do not follow from your premises. Therefore, I cannot accept your conclusions'.

An argument is composed of a number of *premises* and a *conclusion*. The question about the truth of the premises of a scientific argument is an empirical question (except in mathematics and logic); it has to do with what laws or observational statements are true. In this appendix, we shall instead concentrate our discussion on another question, that of whether an argument is *logically valid*; that is, whether one is rationally bound to accept the conclusion if one accepts the premises.

However, before discussing this question it should be mentioned that there are many examples of scientific arguments that do not claim to be logically valid, namely, *inductive arguments*. In such cases, one does not claim that the conclusion logically follows from the premises, but only that the conclusion is probable or believable, given the premises. The question of what kinds of inductive arguments we should, or should not, accept is a source of much controversy in the philosophy of science. Popper claimed that we should never accept an inductive argument, while most other philosophers have claimed that one can accept such arguments under certain conditions. In particular, quite a few have claimed that one can

L.-G. Johansson, *Philosophy of Science for Scientists*, Springer Undergraduate Texts in Philosophy, DOI 10.1007/978-3-319-26551-3

construct rules for determining what degree of credence a certain inductive argument has. I shall not delve deeper into this issue here; rather, I shall turn attention to other cases where the question of *logical validity* is relevant.

The question of whether or not an argument is logically valid is a question about its *logical form*. Validity has nothing to do with the content of an argument, i.e., what the words mean. One should also distinguish the question of whether a *conclusion* is true or not from the question of whether or not an *inference* is valid. A conclusion is true if the premises are true and if the inference is valid, but if any one of the premises is false, the inference can still be valid even though the conclusion is false.

What, then, is the logical form of an argument? To answer this question, we begin with describing the logical form of a single sentence.

Logical Form 1: Sentences

Let us start with a simple assertion, 'The Nile is the longest river in Africa'. This sentence essentially consists of two parts, the subject and the predicate. 'The Nile' is the subject, and 'the longest river in Africa' is the predicate. In high school grammar we identify the predicate with a single word, but in this context we can forget the details and treat entire phrases as predicates. It seems natural to analyse the meaning of the above sentence in the following way: the subject of the sentence *refers* to a certain object, the Nile, and the sentence is true if and only if the Nile has the property that the predicate 'is the longest river in Africa' expresses or describes. But this simple analysis immediately runs into problems in cases where the subject does not exist. A classic example presented by Bertrand Russell is

#1. The present king of France is bald.

One cannot say this sentence is true. Is it, then, false, or meaningless? On the one hand, it seems completely wrong to say that the sentence is meaningless, since its syntax is correct and we understand what the words mean. But is it false? If it is false that the king of France is bald, then it is true that he is not bald, that is, it is true that

#2. The present king of France has hair on his head.

But that is also totally unacceptable. So the sentence is meaningful and well formed, but neither true nor false. Should we give up the principle of bivalence: that every well-formed sentence is either true or false? Most would say 'No!'

In order to solve this problem it is appropriate to introduce a distinction between a sentence's *grammatical form* and its *logical form*. The idea is that the logical form so to speak lies beneath the grammatical form. We can conceive that there exists a certain fundamental proposition with a certain logical structure, which–through various transformations–becomes the sentences we are familiar with in everyday language. In the case of the bald French king, Russell solved the problem by proposing that #1 should be interpreted as a shortened version of the proposition

#3. There now exists one and only one king of France, and this king is bald.

This sentence consists of two separate propositions connected by conjunction 'and'. Notice that #3 can be further reformulated in the following way:

#4. There exists an object x such that x is presently the king of France and x is bald, and if there exists an object y such that y is presently the king of France and is bald, then $x = y$.

This is the logical form of #1, and as such it is quite easy to analyse. Since the sentence is composed of three parts conjoined by two occurrences of the word 'and', each component must be true in order for the entire sentence to be true; and since there is no object x that makes the first part of the conjunction true, then the entire sentence is false. Thus as we expected, the sentence is false, meaningful and completely intelligible. The negation of this false sentence must be true, which is the case, for if we negate #4, we get:

#5. There exists no object x such that x is the present king of France and x is bald.

We have thus fulfilled all the requirements; namely, (i) the sentence, appropriately paraphrased, is meaningful, (ii) it is false and (iii) its negation is true.

There are various other semantic problems, which can be solved by this way of analysing sentences. One such problem is the use of 'nothing' and other indefinite pronouns as subjects. For example, take the sentence

#6. Nothing is sacred.

In normal situations, one would say that a sentence expresses something about what the noun phrase refers to, and if it does not refer to anything, then the sentence is meaningless. But what does 'nothing' refer to? Nothing! In order to maintain the quite reasonable idea that a meaningful assertion consists of a noun phrase and a predicate, and that it is true only if the object to which the noun phrase refers actually has the property that the predicate describes, then it seems that we need to presume that the term 'nothing' refers to a kind of object of thought. However, this is absurd. A simple solution presents itself if we recognize that the logical form of #6 is

#7. There exists no object x such that x is sacred.

Hence we see how greatly variables and the quantifiers 'There exists' and 'For all' help us when we try to describe and understand the semantic aspects of language.

We have now come to a sentence's logical form. The form of a sentence is determined by how it is constructed by means of logical constants and variables. Beside the already introduced constants 'There is', not' and 'and', there are five more, viz., 'all', 'or', 'if…then', and 'if and only if'. These logical constants are often symbolized in the following way:

all	$\forall$
there exists	$\exists$
and	$\wedge$ or &
or	$\vee$
not	$\neg$ or ~
if...then...	$\supset$ or $\rightarrow$
if and only if	$\leftrightarrow$

Logical constants are words whose meaning is completely context-independent. Their meaning can be described by means of introduction and elimination rules. Here is an example illustrating this point.

If we know that sentence A and sentence B are true, then we can introduce the logical constant 'and' according to the following inference rule: (above the line are sentences already accepted and beneath the inferred sentence):

$$\frac{A \quad B}{A \text{ and } B}$$

where we have concluded that 'A and B' is true. (One cannot explain this rule in words without using the word 'and', which is the word we want to explain. Thus there is a point in using structural symbols to display inference forms.) Conversely, if we know that 'A and B' is true, we can eliminate 'and' by the logical rule

$$\frac{A \text{ and } B}{A}$$

i.e., we conclude that 'A' is true, from the premise that 'A and B' is true. Via these two logical rules, we have clearly defined how the word 'and' should be used regarding its insertion in a text, or what conclusions can be drawn from a text in which it is contained. In the same way, one can configure an introduction and elimination rule for all seven logical constants.

Variables

The other component present in #4–#7 is the variable x. This variable functions exactly like a variable in mathematics; that is, it assumes different values in different concrete situations. The difference is that in mathematics variables usually assume numbers as values, whereas logical variables assume objects.

A moment of consideration tells us that pronouns are actually a type of variable in that they do not denote the same objects regardless of context. Instead, they refer according to the circumstances. We can thus say, in general, that pronouns can be replaced with variables such as x, y, or z. This is a great advantage, for constructions that easily become ambiguous in everyday language can be disambiguated when rewritten in their logical form. Consider the following example.

#8. Carl often played tennis with Peter. Most of the time, he lost.

It is not clear whom it was that normally lost, but if we rewrite this sentence in its logical form, the problem can easily be cleared up. The trick is to give the same variable, e.g. x, for 'he' and 'Carl', if we in fact meant that it was Carl who lost most of the time. Thus we get

#9. There exists an object x such that x is Carl and x often played tennis with Peter and x lost most of the time.

Predicates

We have hitherto identified two types of sentence components: logical constants and variables. The third component is the remainder of a sentence, the predicate. I have already mentioned that the word 'predicate' denotes not only individual verbs but also entire expressions, which we use to claim something about the object of discussion. A predicate can be described as an incomplete sentence; start with a complete sentence and substitute one or more expressions that describe or refer to individual things with empty placeholders or variables. In the examples above, the expressions '...is the present king of France', '...is bald' and '... often played tennis with Peter' are predicates. Instead of writing a dotted line to mark an empty space we simply insert a variable, which makes it easier to express. Thus 'x is the present king of France' is the same predicate as '...is the present king of France'. We see here how convenient it is to use variables as placeholders.

One could ask oneself why one should substitute a variable for 'Carl' but not for 'Peter'. The answer is, of course, that one can do so if need be. That is to say, we can either analyse #8 as consisting one singular term 'Carl' and a predicate '...often played tennis with Peter', or as two singular terms 'Carl' and 'Peter' and the *two-place* predicate '...often played tennis with...'. Predicates can thus have one or more empty spaces; we can talk about one-place, two-place, three-place predicates and so on.

Since the purpose of the above analysis is to discover the logical form of a sentence, we are not interested in the content of a sentence. This means that we do not care about what kind of object we are talking about; that is, the objects to which the singular terms refer. Nor are we concerned about the properties we ascribe to them; therefore, it is useful to replace long predicate expressions with placeholders, just as we replace pronouns and logical words with variables and symbols, respectively. The result is maximal lucidity with all informative content removed, thus allowing us to consider the logical form in isolation. This yields a rather compact and perspicuous notation, which simplifies discussion of the form of an argument.

The convention is to replace predicate expressions with upper case letters. In our example #9 above, if we put 'x = Carl' as 'Cx', 'x often played tennis with Peter' as 'Tx' and 'x lost most of the time' as 'Lx' we get

#10. $\exists x Cx \,\&\, Tx \& \, Lx$

which we read as

#11. There exists an x such that x has property C and x has property T and x has property L.

Logical Form 2: Argument

In this appendix, we will not give a complete review of which argument forms are valid and which are invalid; rather, we shall look at some common cases of scientific argument forms as well as some common invalid argument forms.

Some Common Valid Argument Forms

1. Modus Ponens

Perhaps the most common of all argument forms is the following: It is the case that if p is true, then q is true (where p and q are propositions). But we know that p is true (p is the case, it is a fact that p, etc.). Thus it is also the case that q is true.

Using our logical symbols, we can clarify this argument in the following way:

$$\begin{array}{c} p \supset q \\ p \\ \hline q \end{array}$$

Above the solid line we have the premises of the argument, and below it follows the conclusion. I think that everyone would agree that this is a logically valid argument form.

2. Modus Tollens

Example If Carl had travelled to Crete, he would have taken his passport. However, Carl's passport is still at his house. Hence Carl has not travelled to Crete.

In symbolic form, this argument becomes

$$\begin{array}{c} p \supset q \\ \neg q \\ \hline \neg p \end{array}$$

(where p = Carl travelled to Crete, and q = Carl took his passport).

I have here treated ‘Carl did not take his passport’ and ‘Carl’s passport is still at his house’ as logically equivalent. In certain situations, such identification may be unacceptable. A more complete argument is the following.

If Carl has travelled Crete, then he must have taken his passport. However, Carl’s passport is still at his house. If Carl’s passport is at home, then he has not taken his passport with him. Thus Carl has not travelled to Crete. The form of this more complete argument is

1. $p \supset q$ (Assumption)
2. r (Assumption. r=Carl’s passport is at home)
3. $r \supset \neg q$
4. $\neg q$

---

5. $\neg p$

Note that this argument makes use of Modus Ponens in lines 2–4

3. Instantiation

In the discussion of scientific laws and explanations, we often find arguments that can be exemplified by the following: It is a general law that green plants produce oxygen. Now, plankton *p* is a green plant. Therefore, where we find this plankton, we should find the production of oxygen taking place. The sentence ‘all green plants produce oxygen’ has the logical form

For all objects x such that x is a green plant, x produces oxygen.

If we insert the symbols ‘Gx’ for ‘x is a green plant’ and ‘Px’ for ‘x produces oxygen’, then we can formalize this argument in the following way:

$$\frac{\begin{array}{c}\forall x\,(Gx \supset Px)\\ Ga\end{array}}{Pa}$$

This argument is valid. What goes for all objects of a certain kind obviously goes for a particular object of the same kind.

Some Invalid Argument Forms

1. *Confirming the antecedent*

Example (P1) If you are kind, then you will receive Christmas presents from Santa Claus.

(P2) In fact, you do receive Christmas presents from Santa Claus.

C: Therefore, you have been kind.

This is not a logically valid inference since the first premise does not exclude the possibility that one can receive presents without being kind. (In fact, this has been known to happen!) Thus, constructing a counter example is enough for proving an argument being invalid.

The logical form of this invalid argument is:

$$\begin{array}{l} p \supset q \\ q \\ \hline p \end{array}$$

Another example of the same mistake is the following: 'All members of the Left Party want to increase equality and fairness in society. Erik wants to increase equality and fairness is society. Therefore, Erik is a member of the Left Party'.

The form of this argument is (Erik is identified as the referent of the constant *a*):

$$\begin{array}{l} \forall x\,(Gx \supset Px) \\ Pa \\ \hline Ga \end{array}$$

2. *Implication-negation*

Example (P1) If one speaks Swedish fluently, then one was born and raised in Sweden.
(P2) But Ahmed does not speak Swedish fluently.
C Therefore, he was not born in Sweden.

This argument sounds pretty reasonable, especially since the person in question has an Arabic name. However, reasonable or not, the conclusion does not follow from the premises an it is easy to conceive of counterexamples. A study of the logical form soon clarifies this point. The logical form of the premise P1 is

For all x, it is the case that if x speaks Swedish fluently, then x was born and raised in Sweden,

which in symbolic form becomes

$$\forall x\,(Sx \supset Px),$$

where Sx = 'x speaks Swedish fluently' and Px = 'x was born and raised in Sweden'. Since by assumption this is the case for all persons, it is also the case for Ahmed, which we refer to with the letter 'a'. It is thus also the case that

$$Sa \supset Pa.$$

This step can probably be considered as an implicit assumption in the example above. Thus the logical form of the example is the following:

$\forall x\,(Sx \supset Px)$	P1
$Sa \supset Pa$	from P1
$\neg Sa$	P2

$\neg Pa$	C

If this were a logically valid inference, then all arguments with the same form would be such that if the premises are true, then the conclusion is true. It is, however, quite easy to find a counterexample. Consider the following:

If it rains, the ground is wet.
It is not raining.

The ground is not wet.

However, it is entirely possible for the ground to become wet by watering it with a garden hose, even though it has not rained.

It is important to truly understand the 'if…then' construction, which can best done by studying a table of truth-values. Unfortunately, in everyday language use, we use the 'if…then' construction in many different ways, which deserve a little investigation.

The 'if…then' Construction

Conditionals expressed by the 'if…then' constructions are common. An example is 'if the disease is caused by a bacterium, then the disease is cured with penicillin'. Such constructions can be interpreted in, at least, four different ways: (i) material implication (also called 'material conditional'), (ii) causal connection, (iii) derivability or (iv) necessitation.

The interpretation of the 'if…then' construction as *material implication* is based on the extensionality principle, which states that the truth-value of a compound sentence depends *only* upon the truth-values of its components. (This principle can be formulated even for the components of a sentence: a sentence's truth-value does not change if one substitutes one component in the sentence (e.g. a name) for another component with the same extension/reference.) A sentence of the form 'if A, then B' obviously consists of two components, A and B, which are themselves sentences. Thus if we apply the extensionality principle, the truth-value of 'if A, then B' is unequivocally determined by the truth-values of A and B. Combining two sentences into one, we get four combinations of the truth-values true (T) and false (F). The truth-values of 'if A, then B' for each combination are shown in the following table (Table A.1):

Table A.1 The truth table for the material conditional

A	B	If A, then B
T	T	T
T	F	F
F	T	T
F	F	T

Note that if the antecedent is false, then the entire 'if...then' sentence is true independent of the truth-value of the consequent. This fact shocks some people at first, but it is in accord with our use of conditionals. The sentence 'if A, then B' is synonymous with 'if A is true, then B is true'. So the construction 'if....then....' is used for making a conditional statement: we assert the truth of B, conditional on the truth of A, and consequently, if A happens to be false, we do not assert anything about B, it could be true or false.

If we were to require that A cannot be false if B is to be true, then we would have to say 'B is true if and only if A is true'. It is obvious that this is a stronger proposition than 'if A is true, then B is true'. The sentence 'if A, then B' is thus equivalent to the sentence 'it is not the case that A is true and B is false'.

In this interpretation, we have not required that which makes A true also, in some way or other, *brings about, leads to, or causes* B to be true; it suffices that B *in fact* is true, for whatever reasons, when A is true. According to this interpretation, the sentence 'if the moon is made of green cheese, then $7+5=12$' is true!

If we return for a moment to the argument form implication-negation above, we can now clearly see that it cannot be a valid argument form. If we know that the conditional statement is true, then we can certainly eliminate the possibility that A is true and B is false. That is to say, we can exclude the second row of the table. Furthermore, we know that A is false, that is, we eliminate the first row. However, these two pieces of information do not allow us to conclude that B is false, as we still have two alternatives (rows 3 and 4) left, and we do not know whether B is true or false.

Interpretations (ii), (iii) and (iv) represent stronger, non-extensional interpretations of the 'if...then' construction. These interpretations state that the truth-value of a sentence is not only determined by the truth-values of its components; in addition there must exist some kind of relation between the *contents* of the sentences 'A' and 'B'. There are at least three *non-extensional* interpretations of 'if A, then B';

(a) causal connection: A causes B
(b) logical connection: B is derivable from A
(c) modal connection: A necessitates B

(a) Sometimes the most reasonable interpretation of the 'if...then'-construction is that it expresses a causal connection, as in the following sentence: 'if one steps on the gas pedal, the car accelerates'. It is reasonable to interpret such a proposition as if the speaker is taking for granted that there is a causal connection involved, which is so well known that it needs not be explicitly stated.

(b) One can sometimes interpret the 'if...then'-construction as logical derivability as in the following example: 'If all swans are white, then a bird that is not white cannot be swan'. A formal proof is hardly necessary in this case, for everyone can see that the consequence logically follows from the antecedent. Another example is 'if a triangle equilateral, then each of its angles is 60°'. It is easy to see that it is possible to prove the consequence, given the antecedent and the axioms of Euclidian geometry.

(c) The sentence 'if John Doe is a bachelor, then he is an unmarried male' is not a logical truth, but we might think that it is *necessarily true*. But what does it mean to say that a sentence is not merely true, but necessarily true? In this particular case, it appears reasonable to say it is necessarily true on the basis of the definition of a bachelor. However, the obvious objection is that such a definition depends on an established language, which is to say that we usually understand 'bachelor' and 'unmarried male' as being synonymous. This is an historical fact about our language use, and as such is hardly a necessary truth. But then how can a consequence of an historical fact be a necessary truth?

It is not easy to say what it meant by a sentence being necessarily true, if what is meant is something beyond logical truth. The common view nowadays is to explain the meaning of 'p is necessarily true' as that the proposition p is true in all possible worlds that are accessible from our world. This can be formalized so as to satisfy high demands on stringency. But empiricists are not satisfied; fulfilling formal truth conditions is one thing, but as an explanation of the meaning of 'necessary' it is not satisfying; if one wonder about 'necessary' it is hardly a step forward to explain this concept using 'possible worlds' and 'accessibility'; if anything this makes things worse. (But all accept, even empiricists, that the formal semantics in terms of possible worlds shows that talk about necessary truths can be made consistent.)

According to Quine, who has been the staunchest critic of modal talk, there are no other necessary truths than logical truths. Many have protested, arguing that, for example, natural laws as necessary truths that are not logical truths. This is indeed a controversial question, and for my part I have suggested an explanation of the necessity of some natural laws that even a strong critic of modal talk like Quine plausibly would accept (see Sect. 10.7). In any case, the fact remains that a conditional of the form of an 'if...then' sentence is sometimes intended to be interpreted as a necessary truth. Those who claim such an interpretation to be correct are required to explain what they mean by 'necessity'.

According to an influential view, first proposed by Kripke, there really are truths that are metaphysically necessary, such as some identities. This topic is however beyond the scope of this short appendix.

Necessary and Sufficient Conditions

The expressions 'necessary condition' and 'sufficient condition' or their synonyms, are often used. These should not be interpreted any stronger than as variations of the material conditional. In other words, 'A is a sufficient condition for B' means 'If A, then B.' and' 'A is a necessary condition for B' means, 'If B, then A.'. Despite surface appearance, there are no modal commitments in using the term 'necessary condition'.

Further Reading

Priest, G. (2000). *Logic: A very short introduction.* Oxford: Oxford University Press.

Haack, S. (1978). *Philosophy of logics.* Cambridge: Cambridge University Press.

Read, S. (1994). *Thinking about logic: An introduction to the philosophy of logic.* Oxford: Oxford University Press.

Definitions of Some Core Concepts

(N.B. Some of these definitions are highly controversial and some are my own inventions. The reader is urged not to view these definitions as generally agreed upon among philosophers.)

Ad hoc-hypothesis	An auxiliary assumption which (i) is introduced only for saving the tested hypothesis from refutation, and (ii) which do not allow of independent testing.
Analytic sentence	A sentence is analytic if and only if its truth-value only depends on the meaning of its words.
Auxiliary assumption	A statement that (i) is necessary in order to infer an empirical implication from a hypothesis, and which (ii) is not tested in the given situation, but is rather assumed to be true.
Brute fact-social fact	A brute fact is a fact whose existence does not depend on human thoughts, attitudes or knowledge, whereas the existence of social facts does depend on these things.
Conditional Probability	$P(A\|B) = P(AB)/P(B)$
Corroboration (Popper's concept)	An hypothesis is corroborated if and only if has survived attempts to falsify it.
Deduction:	A form of reasoning that is truth-preserving: if the premises in a deductive argument is true, then the conclusion is true.
Empirical Consequence:	A statement that (i) follows from the hypothesis and eventual auxiliary assumptions, and (ii) whose truth, under plausible circumstances, can be determined by observation.

L.-G. Johansson, *Philosophy of Science for Scientists*, Springer Undergraduate Texts in Philosophy, DOI 10.1007/978-3-319-26551-3

Epistemic objective-subjective distinction:	A sentence is epistemically objective if its truth-value is independent of people's opinions, otherwise is it epistemically subjective.
Explanans	That which explains.
Explanandum	That which is being explained.
Extension of a concept	the set of things the concept is true of.
Hypothesis	A statement that (i) we are not entirely certain about, and (ii) which is used as a premise in inferring empirical consequences.
Implicit definition	A complete sentence stated to be true and in which the defined term occur.
Independent events	Two events A and B are independent of one another if and only if P(A and B) = P(A)P(B).
Induction	A form of reasoning that proceeds from a set of sentences to a logically stronger sentence. (This means that even if the premises all are true, the conclusion might be false.)
Intentional	A state of affairs or event is intentional if it has the structure of being an act directed towards an object.
Intension	The meaning of an expression.
Institutional fact	Y is an institutional fact if X counts as Y in contexts C.
Incommensurable	Two theories are incommensurable if there is no common standard by which we can compare them.
INUS-definition of cause	A cause is an Insufficient but Necessary part of an Unnecessary but Sufficient condition for the effect.
Necessary condition	A is a necessary condition for B if and only if the sentence 'if B, then A' is true.
Occasion sentence	A 'here-and-now-sentence' i.e., a sentence describing an aspect of the immediate environment of a speaker and being such that other people being present and who understands the speaker's language immediately can agree on what is said, no matter differences in culture and theoretical beliefs.
Ontological objective-subjective distinction	An event or state of affairs is ontologically subjective if it occurs in an individual person's mind, otherwise ontologically objective.
Propositional knowledge:	Justified, true belief.
Qualitative method	A scientific method is qualitative if and only if it aims at the classification of phenomena with respect to categories containing an explicit, or implicit, intentional component.

Scales

Quotient scale	A measure f for a quantity q is a *quotient scale* if and only if for any other measure g of the same quantity there exists a number $k > 0$ such that $f = kg$
Interval scale	A measure f for a quantity q is an *interval scale* if and only if for any other measure g of the same quantity there exists a number $k > 0$ and a real number r such that $f = kg + r$
Ordinal scale	A measure f for a quantity q is an *ordinal scale* if and only if for any other measure g of the same quantity there exists a strictly increasing function j such that $g = j(f)$
Nominal scale	A sorting of items into categories without any measure being applied is a nominal scale.
Synthetic sentence	A sentence is synthetic if its truth-value not only depends of its meaning but also on empirical facts.
Sufficient condition	A is a sufficient condition for B if and only if the sentence 'if A, then B' is true.
Theory	A set of statements whose relations are explicitly stated.
Value-free	A theory is value-free if it does not contain any value-statements.
Value-laden	A theory is value-laden if pursuing the activity of exploring the theory is motivated by certain values.
Verifiability criterion	A sentence is meaningful if and only if it, or its negation, can be verified.

Literature

Achinstein, P. (1981). Can there be a model for explanation?. *Theory and Decision, 13*, 201–227.
Achinstein, P. (1983). *The nature of explanation*. New York: Oxford University Press.
Achinstein, P. (1984). The pragmatic character of explanation. *PSA Proceedings of the Biennial Meeting of the Philosophy of Science Association*, Vol. 1984, Volume two: Symposia and invited papers (1984) (pp. 275–292). Chicago: University of Chicago Press.
Alvesson, M., & Sköldberg, K. (2008). *Tolkning och reflektion. Vetenskapsfilosofi och kvalitativ metod*. Lund: Studentlitteratur.
APA Ethical Code. http://www.apa.org/ethics/code/principles.pdf
Aristotle. (1984). *Metaphysics* (trans: Ross, W.D.). https://ebooks.adelaide.edu.au/a/aristotle/metaphysics/
Aristotle. (1966). *Posterior analytics* (trans: Mure, G.R.G.). https://ebooks.adelaide.edu.au/a/aristotle/a8poa/index.html.
Armstrong, D. (1983). *What is a law of nature?*. Cambridge: Cambridge University Press.
Berger, P., & Luckman, T. (1967). *The social construction of reality*. New York: Anchor Books.
Bird, A. (2007). *Nature's metaphysics, laws and properties*. Oxford: Oxford University Press.
Box-Steffensmeier, J. M., Brady, H. E., & Collier, D. (Eds.). (2008). *The oxford handbook of political methodology*. Oxford: Oxford University Press.
Broad, W. (1985). *Betrayers of the truth: Fraud and deceit in science*. Oxford: Oxford University Press.
Brown, J. (1994). *Smoke and mirrors. How science reflects reality*. London: Routledge.
Bruun, H. H. (2007). *Science, values and politics in Max Weber's methodology: New expanded edition*. Aldershot: Ashgate.
Carroll, J. (1994). *Laws of nature*. Cambridge: Cambridge University Press.
Cartwright, N. (1983). *How the laws of physics lie*. Oxford: Clarendon Press.
Cartwright, N. (1989). *Nature's capacities and their measurement*. Oxford: Clarendon Press.
Cartwright, N. (1999). *The dappled world. A study of the boundaries of science*. Cambridge: Cambridge Univ. Press.
Chalmers, D. (1996). *The conscious mind*. Oxford: Oxford University Press.
Chesterman, A. (2008). The status of interpretive hypotheses. In G. Hansen, A. Chesterman, & H. Gerzymisch-Arbogast (Eds.), *Efforts and models in interpreting and translation research: A tribute to Daniel Gile* (pp. 49–61). Philadelphia: John Benjamins.
Churchland, P. M. (1988). *Matter and consciousness*. Cambridge, MA: MIT Press.
Churchland, P. M. (1995). *The engine of reason, the seat of the soul*. Cambridge, MA: MIT Press.
Churchland, P. S. (2013). *Touching a nerve. Our brains, our selves*. New York: W.W. Norton &Co.

L.-G. Johansson, *Philosophy of Science for Scientists*, Springer Undergraduate Texts in Philosophy, DOI 10.1007/978-3-319-26551-3

Collingwood, R. G. (1994). *The idea of history* (Revised ed.). Oxford: Oxford University Press.
Confucius. (500 BC/1994). *Analects*. http://classics.mit.edu/Confucius/analects.html
Cresswell, J. W. (1998). *Qualitative inquiry and research design*. Thousand Oaks: Sage Publications.
Davidson, D. (1963). *Actions, reason, causes*. In his (1980), pp. 3–20.
Davidson, D. (1980). *Essays on actions and events*. Oxford: Clarendon Press.
Davidson, D. (1984). *Inquiries into truth and interpretation*. Oxford: Clarendon Press.
Dennett, D. (1981). *Brainstorms*. Cambridge, MA: MIT Press.
Dijkman, R., Dumas, M., & Garcia-Banuelos, L. (2009b). Graph matching algorithms for business process model similarity search. In U. Dayal et al. (Eds.), *Business process management*. Berlin: Springer.
Dummett, M. (1993). *Origins of analytical philosophy*. London: Duckworth.
Earman, J. (2002). *Laws, symmetry, and symmetry breaking; invariance, conservation principles, and objectivity*?. philsci-archive.pitt.edu/878/1/PSA2002.pdf
Ellis, B. (1968). *Basic concepts of measurement*. Cambridge: Cambridge University Press.
Elster, J. (1994). Functional explanation: In social science Chapter 25. In Martin and McIntyre (1994).
Englund, P. (2000). *Den oövervinnerlige: om den svenska stormaktstiden och en man i dess mitt*. Stockholm: Atlantis.
Fay, B. (1996). *Contemporary philosophy of social science*. Oxford: Blackwell.
Feyerabend, P. (1988). *Against method*. London: Verso.
Floridi, L. (2011). *The philosophy of information*. Oxford: Oxford University Press.
Friedman, M. (1974). Explanation and scientific understanding. *Journal of Philosophy, 71*, 5–19.
Friedman, M., & Rosenman, R. H. (1974). *Type A behavior and your heart*. New York: Knopf.
Gibson. (2004). *Quintessence. Basic readings from the philosophy of W.V. Quine*. Cambridge, MA: Belnap Press.
Giere, R. (1999). *Science without laws*. Chicago: Chicago University Press.
Glaser, B. G., & Strauss, A. L. (1967). *The discovery of grounded theory: Strategies for qualitative research*. New York: Aldine de Gruyter.
Glassner, B., & Moreno, J. D. (Eds.). (1989). *The qualitative-quantitative distinction in the social sciences*. Dordrecht: Kluwer.
Goodman, D. C., & Russell, C. A. (1991). *The rise of scientific Europe 1500–1800*. London: Hodder & Stoughton.
Grant, E. (1996). *The foundations of modern science in the middle ages: Their religious, institutional, and intellectual contexts*. Cambridge: Cambridge University Press.
Haack, S. (1978). *Philosophy of logics*. Cambridge: Cambridge University Press.
Hacking, I. (1999). *Social construction of what?* Cambridge, MA: Harvard University Press.
Hannam, J. (2011). *The genesis of science: How the Christian middle ages launched the scientific revolution*. Washington, DC: Regnery Publishing.
Hanson, N. R. (1965). *Patterns of discovery: An inquiry into the conceptual foundations of science*. Cambridge: Cambridge U. P.
Harding, S. (1991). *Whose science? Whose knowledge?* Press: Cornell University.
Hempel, C. (1965). *Aspects of scientific explanation and other essays in the philosophy of science*. New York: Free Press.
Hesslow, G. (1984). What is a genetic disease. On the relative importance of causes. In Lindahl & Nordenfelt (1984).
Hillel-Ruben, D. (Ed.). (1993). *Explanation. Oxford readings in philosophy*. Oxford: Oxford University Press.
Hippocrates. (2000). *On the sacred disease* (trans: Francis, A.). http://classics.mit.edu/Hippocrates/sacred.html
Hitchcock, C. (2001). The intransitivity of causation revealed in equations and graphs. *Journal of Philosophy, 98*, 273–99.
Hospers, J. (1986). *Introduction to philosophical analysis*. London: Routledge.

Johansson, L.-G. (2007). *Interpreting quantum mechanics. A realist view in Schrödinger's vein*. Aldershot: Ashgate.
Johnstone, B. (2000). *Qualitative methods in sociolinguistics*. New York: Oxford University Press.
Kahneman, D. (2011). *Thinking, fast and slow*. New York: Farrar, Straus and Giroux.
Kant, I. (1781/1787) *Critique of pure reason* (trans: Meiklejohn, J.M.D.). https://ebooks.adelaide.edu.au/k/kant/immanuel/k16p/index.html
Kim, J. (1998). *Philosophy of mind*. Boulder: Westview Press.
Kirk, J., & Miller, M. L. (1986). *Reliability and validity in qualitative research*. Newbury Park: Sage Publications.
Kitcher, P. (1981). Explanatory unification. *Philosophy of Science, 48*, 507–31.
Kitcher, P., & Salmon, W. (1987). Van Fraassen on explanation. *Journal of Philosophy, 84*, 315–330.
Kruschke, J. (2014). *Doing Bayesian data analysis, second edition: A tutorial with R, JAGS, and Stan*. Burlington: Academic.
Kuhn, T. S. (1962). *The structure of scientific revolutions*. Chicago: University of Chicago Press.
Kuhn, T. S. (2000). *The road since structure: Philosophical essays, 1970–1993*. Chicago: University of Chicago Press.
Ladyman, J., & Ross, D. (2007). *Every thing must go: Metaphysics naturalized*. Oxford: Oxford University Press.
Lakatos, & Musgrave (Eds.). (1970). *Criticism and the growth of knowledge*. Cambridge: Cambridge University Press.
Lange, M. (2009). *Laws and lawmakers*. Oxford: Oxford University Press.
Lindahl, B. I. B., & Nordenfelt, L. (Eds.). (1984). *Health, disease and causal explanation in medicine*. Dordrecht: Reidel.
Lindberg, D. (1992). *The beginnings of western science*. Chicago: University of Chicago Press.
Longino, H. (1990). *Science as social knowledge*. Princeton: Princeton University Press.
Longino, H. E. (2002). *The fate of knowledge*. Princeton: Princeton University Press.
Mackie, J. (1974). *The cement of universe*. Oxford: Clarendon Press.
Martin, M., & McIntyre, L. C. (Eds.). (1994). *Readings in the philosophy of social sciences*. Cambridge: Cambridge University Press.
Moore, & McCabe. (2005/2009/2012/2014). *Introduction to the practice of statistics*. New York: W.H. Freeman and Co.
Merton, R. (1973). The normative structure of science. In *The sociology of science* (pp. 267–278). Chicago: University of Chicago Press.
Miettinen, O. (1985). *Theoretical epidemiology*. New York: Wiley.
Mumford, S. (2004). *Laws of nature*. London: Routledge.
Newton, I. (1999). *The Principia: Mathematical principles of natural philosophy*. Berkeley: University of California Press. First published 1687.
Newton-Smith, W. (1981). *Rationality of science*. London: Routledge.
Niiniluoto, I. (1987). *Truthlikeness*. Dordrecht: Reidel.
Pearl, J. (2000). *Causality: Models, reasoning, inference*. Cambridge: Cambridge University Press.
Pitt, W. (Ed.). (1988). *Theories of explanation*. Oxford: Oxford University Press.
Popper, K. (1992 [1959]). *The logic of scientific discovery*. London: Routledge.
Priest, G. (2000). *Logic: A very short introduction*. Oxford: Oxford University Press.
Putnam, H. (1960). *Minds and machines*, reprinted in Putnam 1975, pp. 362–385.
Putnam, H. (1967). *The nature of mental states*, reprinted in Putnam 1975, pp. 429–440.
Putnam, H. (1975). *Mind, language, and reality*. Cambridge: Cambridge University Press.
Putnam, H. (2002). *The collapse of the fact/value distinction and other essays*. Cambridge, MA: Harvard University Press.
Quine, W. V. O. (1953a). *From a logical point of view*. Cambridge, MA: Harvard University Press.
Quine, W. V. O. (1953b). *Three grades of modal involvement* (pp. 158–176). In his (1976).
Quine, W. V. O. (1960). *Word and object*. Cambridge, MA: MIT Press.

Quine, W. V. O. (1969). Epistemology naturalized. In his *Ontological relativity and other essays* (pp. 69–90). New York: Columbia University Press.

Quine, W. V. O. (1976). *The ways of paradox and other essays* (2nd ed.). Cambridge, MA: Harvard University Press.

Quine, W. V. O. (1981). *Theories and things*. Cambridge, MA: Harvard University Press.

Quine, W. V. O. (1990). *Pursuit of truth*. Cambridge, MA: Harvard University Press.

Quine, W. V. O. (1993). Praise of observation sentences. In D. Föllesdahl & D. B. Quine (Eds.), *Confessions of a confirmed extensionalist and other essays* (pp. 409–419). Cambridge, MA: Harvard University Press.

Quine, W. V. O. (1995). *From stimulus to science*. Cambridge, MA: Harvard University Press.

Quine, W. V. O. (2004). Two dogmas in retrospect. In Gibson (Ed.), *Quintessence. Basic readings from the philosophy of W.V. Quine*. Cambridge, MA: Belnap Press.

Railton, P. (1978). A deductive-nomological model of probabilistic Explanation. *Philosophy of Science, 45*, 206–226.

Read, S. (1994). *Thinking about logic: An introduction to the philosophy of logic*. Oxford: Oxford University Press.

Redi, F. (1909). *Experiments on the generation of insects*. Chicago: Open Court. First published 1668.

Reichenbach, H., & Reichenbach, M. (1999[1956]). *The direction of time*. Mineola, New York: Dover.

Ricoeur, P. (1981). *Hermeneutics and the human sciences: Essays on language, action and interpretation*. In R. Paul (Ed.), (trans: and introd. by Thompson, J.B.). Cambridge: Cambridge University Press.

Rogers, E. M. (1977[1960]). *Physics for the inquiring mind: The methods, nature, and philosophy of physical science*. 12. pr. Princeton: Princeton University Press.

Russell, B. (1961). *History of Western philosophy*. London: Allen & Unwin.

Salmon, W. (1984). *Scientific explanation and the causal structure of the world*. Princeton: Princeton University Press.

Salmon, W. (1998). *Causality and explanation*. Oxford: Oxford University Press.

Searle, J. (1992). *The rediscovery of mind*. Cambridge, MA: MIT Press.

Searle, J. (1996). *The construction of social reality*. London: Penguin.

Shannon, C. E., & Weaver, W. (1949). *The mathematical theory of communication*. Urbana: University of Illinois Press.

Singh, S., & Ernst, E. (2009[2008]). *Trick or treatment?: Alternative medicine on trial*. London: Corgi.

Sousa, E., & Tooley, M. (Eds.). (1993). *Causation. Oxford readings in philosophy*. Oxford: Oxford University Press.

Stoerig, P. (1996). Varieties of vision: From blind responses to conscious recognition. *Trends in Neurosciences, 19*(9), 401–406.

Suárez, M. (2003). Scientific representation: Against similarity and isomorphism. *International Studies in the Philosophy of Science, 17*, 225–244.

Suppes, P. (1993). *Models and methods in the philosophy of science: Selected essays*. Dordrecht: Kluwer.

Suppes, P. (2002). *Representation and invariance of scientific structures*. Stanford: CSLI Publications.

Suppes, P., & Zinnes, J. L. (1963). Basic measurement theory. In R. D. Luce, R. R. Bush, & E. H. Galanter (Eds.), *Handbook of mathematical psychology* (Vol. 1, pp. 3–76). New York: Wiley.

Taylor, A. J. P. (1969). *War by time-table*. London: McDonald &Co Publishers.

Thucydides. (432 BC). *The history of the Peleponesian War* (trans: Crawley, R.). http://classics.mit.edu/Thucydides/pelopwar.2.second.html

Tosh, J., & Lang, S. (2006). *The pursuit of history: Aims, methods and new directions in the study of modern history* (4th ed.). Harlow: Pearson Education.

van Fraassen, B. (1980). *The scientific image*. Oxford: Clarendon Press.

van Fraassen, B. (1989). *Laws and symmetry*. Oxford: Clarendon.
Villius, E., & Villius, H. (1966). *Fallet Raoul Wallenberg*. Gebers: Stockholm.
Wang, Z., Bovik, A. C., Sheikh, H. R., & Simoncelli, E. P. (2004). Image quality assessment: From error visibility to structural similarity. *IEEE Transactions on Image Processing, 13*(4), 600–612.
Watson, J. B. (1913b). Psychology as the behaviourist views it. *Psychological Review, 20*, 158–177.
Weber, M. (2011). *Methodology of social sciences*. New Brunswick: Transaction Publishers.
Williams, M. (2001). *Problems of knowledge. A critical introduction to epistemology*. Oxford: Oxford University Press.
Wiseman, R. (2011). *Paranormality*. London, Basingstoke and Oxford: MacMillan.
Wittgenstein, L. (1953). *Philosophical investigations*. Oxford: Blackwell.
Woodward, J. (2003). *Making things happen: A theory of causal explanation*. Oxford: Oxford University Press.

Index

L.-G. Johansson, *Philosophy of Science for Scientists*, Springer Undergraduate Texts in Philosophy, DOI 10.1007/978-3-319-26551-3

Printed in the United States
By Bookmasters